# Vaccines
## in the
# Global Era

How to Deal Safely and Effectively
with the Pandemics of Our Time

# Vaccines
## in the
# Global Era

How to Deal Safely and Effectively
with the Pandemics of Our Time

## Rino Rappuoli
*GlaxoSmithKline Vaccines, Italy*

## Lisa Vozza
*AIRC Foundation for Cancer Research, Italy*

Translated from Italian by **Lisa Vozza**
Revised and Edited by **Lucy A. Lennon**

**World Scientific**

NEW JERSEY · LONDON · SINGAPORE · BEIJING · SHANGHAI · HONG KONG · TAIPEI · CHENNAI · TOKYO

*Published by*

World Scientific Publishing Europe Ltd.

57 Shelton Street, Covent Garden, London WC2H 9HE

*Head office:* 5 Toh Tuck Link, Singapore 596224

*USA office:* 27 Warren Street, Suite 401-402, Hackensack, NJ 07601

**Library of Congress Cataloging-in-Publication Data**
Names: Rappuoli, Rino, author. | Vozza, Lisa, author, translator. | Lennon, Lucy A., editor.
Title: Vaccines in the global era : how to deal safely and effectively with the pandemics of our time /
    Rino Rappuoli, GlaxoSmithKline Vaccines, Italy, Lisa Vozza, AIRC Foundation for Cancer
    Research, Italy ; translated from Italian by Lisa Vozza ; revised and edited by Lucy A. Lennon.
Other titles: Vaccini dell'era globale. English
Description: [Second edition] | New Jersey : World Scientific, [2022] |
    Translation of: I vaccini dell'era globale : come affrontare con sicurezza ed efficacia le
    pandemie del nostro tempo. | Includes bibliographical references and index.
Identifiers: LCCN 2022000344 | ISBN 9781800611931 (hardcover) |
    ISBN 9781800612044 (paperback) | ISBN 9781800611948 (ebook for institutions) |
    ISBN 9781800611955 (ebook for individuals)
Subjects: LCSH: Vaccines. | Vaccination. | Epidemics. | Epidemiology.
Classification: LCC QR189 .R364 2022 | DDC 615.3/72--dc23/eng/20220211
LC record available at https://lccn.loc.gov/2022000344

**British Library Cataloguing-in-Publication Data**
A catalogue record for this book is available from the British Library.

For any available supplementary material, please visit
https://www.worldscientific.com/worldscibooks/10.1142/Q0352#t=suppl

Desk Editors: Jayanthi Muthuswamy/Shi Ying Koe

Typeset by Stallion Press
Email: enquiries@stallionpress.com

# About the Authors

**Rino Rappuoli** is an Italian microbiologist. He was awarded the Gold Medal of Merit for Public Health in Italy in 2005 and is among the highest world authorities on vaccines. He is currently Scientific Director and Head of External Research and Development at GlaxoSmithKline Vaccines (GSK). Previously, he was a visiting scientist at the Rockefeller University in New York and at Harvard Medical School in Boston; he has also served as Director of Vaccine Research at Sclavo, Chiron Corporation, and Novartis Vaccines. Among his many achievements, the developments of the first recombinant bacterial vaccine against pertussis, of the MENJUGATE(R) conjugate vaccine against meningococcal-C disease, and of BEXERO against meningococcal-B disease are prominent. A member of the American National Academy of Sciences and other prestigious scientific societies, in 2016, he was elected as a Foreign Member of the Royal Society. He has received countless prizes and awards, the latest of which was the Robert Koch Medal in 2019.

**Lisa Vozza** is a well-known Italian science writer, trained in biology, and Chief Scientific Officer at AIRC Foundation for Cancer Research. Her books include the following: "Nella mente degli altri" (*In the Mind of Others*, Zanichelli, 2007) with Giacomo Rizzolatti; "I vaccini dell'era globale" (*Vaccines in the Global Era*, Zanichelli, 2009, Galileo literary prize 2010) with Rino Rappuoli; and "Come nascono le medicine" (*How Medicines Are Born*, World Scientific, UK, 2017; The Commercial Press, China, in press) with Maurizio D'Incalci. She is also the author of over 230 articles on the blog, "Biologia e dintorni,"

hosted on the Zanichelli website. She is often invited to present her books and writings in schools, bookshops, and other venues, and to teach science writing. Previously, she was the editor of two series of popular science books, *Chiavi di lettura* and *Mestieri della Scienza*, by Zanichelli, and she worked for the European editions of *Scientific American*. In parallel to her work as science writer, she in charge of the peer review process at AIRC, the major cancer charity in Italy: each year thousands of cancer research proposals are evaluated by a select group of 600 international experts, in line with the standards and practices of the most advanced research charities.

# Contents

# Introduction

Before the first Italian edition of this book was published in 2009, the science of vaccines was mostly the concern of experts and specialists, with an almost threadbare popular science section in bookstores. This void needed to be filled, particularly as the available scientific information was excellent.

Thanks to the genomic revolution, older vaccines were replaced with preparations of greater effectiveness and safety. These newly conceived vaccines were produced against diseases that had previously been unpreventable. The public needed to be informed about these new advances in health while being reassured of the vaccines' safety, particularly when considering the countless unfounded reports on the alleged dangers of immunization.

The thirteen years that have passed since that first edition feel like an eternity given the pace of technological progress in the fields of biology and medicine. More than one technological revolution has taken place, further accelerated by the developments prompted by the COVID-19 pandemic. We needed to update our little contribution to modern popular culture about vaccines.

For the first time in history, several brand-new vaccines, safe and effective against a completely new virus, were conceived, tested, and produced in a record time of eleven months. If we have any hope of finding a way out of the current health crisis, it will be thanks to these preparations.

Over the last twenty years, three epidemics have been attributable to a coronavirus: after Severe Acute Respiratory Syndrome (SARS) and Middle East Respiratory Syndrome (MERS), COVID-19 is by far the most devastating. In the absence of vaccines and therapies, we have been forced to coexist with the

COVID-19 as in the days of the "Spanish" flu of 1918: frightened and isolated, with the sound of ambulance sirens as a background to the unfolding economic and social havoc that the contagion inflicted to people almost everywhere around the world. The pandemic has also laid bare the ecological and environmental conditions that have favored the emergence of Severe Acute Respiratory Syndrome Coronavirus 2 (SARS-CoV-2) from an animal reservoir, brutally exposing the vulnerability of our way of life.

While vaccinologists continue to generate ever more precise and effective preparations, the attitude of the people towards vaccines appears to be moving in the opposite direction, at least among a vocal minority of the world population. A dangerous drop in adherence to routine infant vaccines has allowed the reemergence of preventable diseases, such as measles. Outbreaks of vaccine-preventable diseases, resulting from families not vaccinating their children, have forced public health authorities in many countries to reinstate laws, making vaccinations compulsory. This tough but inevitable decision was made to stem an emergency that would not exist were the public better informed, more aware of the risks of infectious diseases, and more capable of discriminating between fake news and trustworthy sources.

The technologies developed against COVID-19, such as vaccines based on messenger RNA (mRNA) and adenoviral vectors, equip us with a valuable legacy for the future. They will be a crucial weapon in the battle against antimicrobial resistance: a slow but inexorable pandemic in the making, which, if not addressed, could render an arm scratch or a tooth extraction lethal.

The vaccines of the future will also improve the quality of life for people who have reached an age unheard of until half a century ago, thanks also to the immunizations received in childhood. Recent advances in immunology are helping to cure several diseases that become more prevalent with age, such as cancer.

However, the world will not free itself easily from the threat of infectious diseases if we fail to immunize the populations of those countries that cannot afford vaccines. We will also have to create vaccines for communicable diseases that are increasingly widespread, not only in the poorest countries, but also in the populous slums of many of the richest cities in the world. Great progress has been made in these domains, thanks also to the innovations and funds made available by organizations such as Gavi, the Vaccine Alliance, and charities such as the Bill & Melinda Gates Foundation.

Vaccines are an extraordinary resource against infectious diseases: they are cheap, they save countless lives, and they are more effective than the best drugs. Let us make the best use of them, for the health of the greatest number of people and animals, who, with us, share life on our beautiful planet, so greatly in need of care.

# 1
# Old and New Epidemics

I t is early January 2020. The Chinese Center for Disease Control and Prevention (CDC) publishes on its website the genetic sequence of a previously unknown virus. In the Wuhan area, the virus is causing a severe form of pneumonia never before seen by physicians. Just three weeks later, the epidemic originating from the new virus, which has in the meantime been called SARS-CoV-2, has reached worrying proportions. The Chinese government places around thirty million people in the city and province of Wuhan into a severe lockdown and within days has established extra hospitals to cater for this emergency. While people in China are being strictly quarantined and isolated for approximately three months, the virus has already left the country, arriving first in Europe, then in the United States, and then in the rest of the world. In early March 2020, the World Health Organization (WHO) announces that COVID-19 (this is the name of the viral disease) has reached pandemic proportions. In other words, no country has been spared.

According to the British immunologist Sir Peter Medawar, viruses are "pieces of bad news wrapped in protein." Along with bacteria and parasites, trillions of these very tiny organisms, invisible to the naked eye, have accompanied our existence since the origin of our species. They were present well before us, and can infect all kinds of microorganisms, plants, and animals. Many are harmless, and only a few hundred have the ability to infect us. The microbes that we know in some detail are in the order of tens of thousands. Furthermore, they are unlikely to give up wanting to be our unwelcome guests anytime soon, as they hop from one body to another.

Yet if so many individuals belonging to our species have somehow survived the many epidemics in history it is because we have evolved a powerful immune system that is also capable of learning. Yes, it is powerful and it can learn a lot, but it is not foolproof, especially when facing microbes it has never encountered before. Our knowledge and ingenuity have helped us develop weapons that strengthen our natural defenses. From this perspective, vaccines are by far the most effective tool, because they prevent both the spread of diseases and the germs that cause them for an entire population. Some infectious diseases can be managed by drugs, which cannot prevent an infection, but can at least treat it. When we have neither vaccines nor drugs, we are left with containment measures: hygiene, quarantine, and isolation.

Globally, the deaths from COVID-19 have exceeded 5.9 million and the estimated number of infected people is above 433 million, as we finish writing this book in February 2022. In almost every place it has spread to, the virus created a sudden and immediate health emergency, followed by an economic and social crisis. As intensive care units were filled beyond capacity and most collective activities stopped, sudden poverty hit almost every person lacking sufficient savings to endure the crisis, with most jobs negatively affected by these new and unexpected conditions.

Unlike those who preceded us, we have had the privilege of living very long periods of time protected from serious epidemics. When viewing the historical perspective, such a privilege is unprecedented. Yet history repeats itself. Today still, the destructive power of infectious diseases can be equal to that of infections that we had hoped would never return, especially when we are faced with a new virus, unknown to the immune system of the entire world population, and for which there are no vaccines or therapies available.

## The greatest existing monument to the destructive force of infectious diseases

Let us take a journey back in time, to medieval Siena in 1348. In this little town in central Italy, an immense construction site is at work: the largest Gothic cathedral in the world is being built. At the end of April, the air is still cool, but hundreds of workers are scrambling up the high scaffolding despite the morning breeze, working tirelessly, like industrious ants. The most famous artists from all over Europe have converged to the rich and beautiful Siena. They are competing to leave their individual mark on the forms of the evolving steeples and chapiters, and in the fabulous paintings that will adorn the vaults and walls.

The new cathedral is being grafted onto the old, adjacent building, now too small to accommodate a population that has grown out of proportion thanks to a booming economy. However, the city's ambitions and its project of expansion are not limited to simply enlarging the ancient church. The future cathedral must eloquently signify the power that Siena is enjoying, which is comparable to modern-day Shanghai. It must also be worthy of the sumptuous palaces that bankers and traders have erected in the city's *contrade*, a lasting symbol of the great fortunes accumulated in just a few years.

Now let us jump ahead to July 1348. Only three months have passed, but the city is unrecognizable. Immersed in the summer heat, the streets are deserted and silent; the shops have been closed for weeks and it is almost impossible to find food. Not a fly buzzes in Piazza Duomo; the construction site is empty. The reason for Siena's silence is neither a violent earthquake nor an army of barbarians, but something much more imperceptible. A small bacterium from the East, *Yersinia pestis*, has reached the Tuscan cities in midspring, sowing death and despair over the next few months.

It is difficult to avoid contagion by the Black Death epidemic ravaging Europe, Asia, and Africa between 1346 and 1353. The disease acts swiftly, generally taking no more than three days from the very first symptoms to death. Siena's mass graves are overflowing and there is no space left to bury the corpses. The few who have escaped contagion have quickly abandoned the city, which remains prey to mice and rats. After less than one hundred days, only one in two Sienese individuals survive, or one in three according to other estimates.

By October the epidemic has finally subsided, but the city is exhausted and depopulated, and the few survivors are unable to ensure everyday operations. No councilors are left to govern the city and many doctors, soldiers, traders, barbers, carpenters, and peasants have died. No single family, profession, or social category has been spared from the ravages of the disease. The economic, social, and cultural disruption is devastating; Siena will never regain the splendor it boasted just a few months earlier, and is now destined to be crushed by competition from other cities that would replace it in commerce, talent, and wealth.

There is no more money, and no artists available to build the naves and spires of the largest cathedral in Europe. What would be the use of such large premises, you may ask, for a population that has been halved or even reduced to a third by the epidemic? Construction is interrupted, never to be resumed. Today, only a bare wall and the frames of a few windows hint of that dream of a magnificent cathedral that should have defied the force of gravity (Figure 1). The unfinished wall is nothing in comparison to the Gothic cathedrals of Northern Europe that the Sienese had taken as a benchmark. That wall and those unfinished windows are perhaps the greatest existing monument to the destructive force of infectious diseases, with their ability to wipe out a thriving economy and an exuberant civilization in a few days.

Figure 1: An unfinished wall of what might have been the new Cathedral of Siena, had the Black Death not destroyed this city and its population at the apex of its success in 1348 (photo by Giorgio Corsi).

The tale of the Sienese plague is just one episode in the Black Death pandemic that, with its recurring waves, may have killed a third of the world population of the time. It remains a highly eloquent episode of the destructive power of microbes.

## Pandemics in history and today

Going further back in time, what is sometimes referred to as the first pandemic in history, appears to have originated in the city of Pelusium, near modern-day Port Said in northeastern Egypt, in 541 CE. According to the historian Procopius, who was a living witness of the period, the "plague" spread initially both to the west, towards Alexandria, and to the east, in the direction of Palestine. It then continued its course. According to Procopius, the contagion seemed to move almost consciously, "as if fearing lest some corner of the earth might escape it."

Diseases such as measles and smallpox, probably introduced to the Americas by the Spanish expedition led by Hernán Cortés in 1519, contributed to the disappearance of the Aztec civilization.

In 1918, the "Spanish" flu spread rapidly from one continent to another, perhaps killing more than forty million people in less than a year. Its death toll was likely higher than for the First World War, which had just ended. But the consequences of the "Spanish" flu go well beyond the number of deaths. It resulted in a sharp decline—ten or so years—in life expectancy (Figure 2).

Infectious diseases such as the plague of 1348, the "Spanish" flu of 1918, or COVID-19 in 2020, have the impetus of a tsunami: swooping quickly over a country, a city, a nation; spreading swiftly across continents only to at times disappear just as quickly, leaving behind piles of human rubble. Other *pathogens* (a technical term that means "carriers of diseases") spread more slowly among people, but they are no less lethal for it. A tragic example, sadly still ongoing, is the acquired immunodeficiency syndrome (AIDS) epidemic that has been raging all over the world for more than thirty-five years, especially in Africa. The infection, caused by the human immunodeficiency virus (HIV),

Life expectancy, United States, 1880–2019

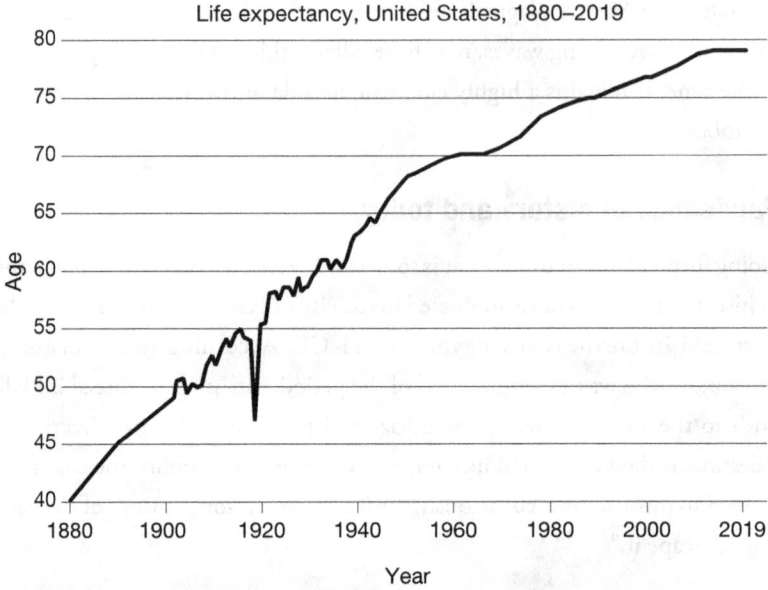

Figure 2: Life expectancy in the United States from 1880 to 2019. The deep indentation just before 1920 was caused by the Spanish flu of 1918–1919 (adapted from *Our World in Data*).

insinuates itself subtly into individuals, causing a symptom-free infection that can last for more than a decade before manifesting itself as a full-blown disease. This can happen because HIV is capable of applying a sort of "silencer" to itself, allowing the virus to multiply in the infected organism without disclosing itself to its host. Just as silently, it spreads across an unprotected population.

The result has been catastrophic: around 36 million people have died and 73 million have been infected worldwide since the beginning of the epidemic, according to WHO estimates. Today still, the number of deaths and infections caused by HIV continues to rise. In 2018, in Botswana, one of the most affected countries, more than 20 percent of the adult population (ages 15 to 49) was infected. For a while the life expectancy curve in sub-Saharan Africa has stopped growing dramatically in the last 40 years (Figure 3), mainly due to AIDS, while it has risen almost unequivocally in the rest of the world.

Life expectancy, Sub-saharan Africa, 1960–2018

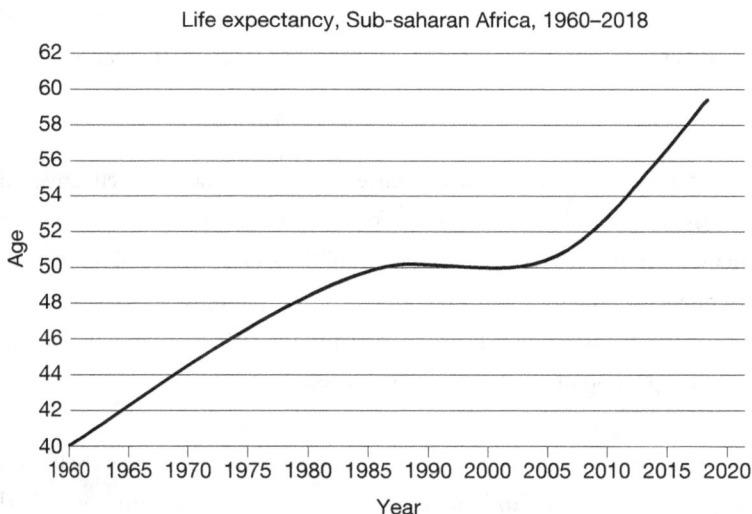

Figure 3: Life expectancy stopped growing in sub-Saharan Africa in the last forty years, mainly as a consequence of AIDS (World Bank).

Numbers can tell us a lot, but not everything. In Africa, millions of orphans are growing up without the knowledge, teachings, and traditions normally transmitted to them by the previous generation, most of whom have been seriously affected by the disease. An entire civilization, with its social, cultural, and economic life, risks disappearing, at least as we know it.

## We can defend ourselves from microbes

The picture is undoubtedly bleak, the threat of infectious diseases daunting. However, we should bear in mind that by now our species' immune response has been tested against microbes for several hundred thousand years. If there are almost eight billion people on Earth today, and if the human population keeps growing, it is because our history is not only dotted with terrible epidemics, but also with sound reasons for being optimistic about our defense capabilities. And vaccines provide an extraordinary reason for optimism, as we will try to explain in this book.

Many of us carry the memory of the vaccinations we received as children in our bodies. Those injections, administered to several billion people from the latter part of the last century onwards, have eradicated widespread and deadly diseases such as smallpox and polio, which only the elderly among us may perhaps remember.

Of the tools available to medical science, vaccines are far more effective than the most effective medicines. In this century alone, spanning at least five or six human generations, they have prevented millions of diseases and their terrible consequences, including but not limited to the richest countries that could afford them. Some scholars have estimated that vaccines are responsible for a significant portion of our increased life expectancy.

The "secret" of the effectiveness of vaccines against diseases lies in their ability to exploit one of the fundamental mechanisms of immunity, which the American philosopher of science, Daniel Dennett, has described with an effective metaphor in his book *Consciousness Explained* (Dennett, 1997):

> "Me against the world"—this distinction between everything on the inside of a close boundary and everything in the external world—is at the heart of all biological processes, not just ingestion and excretion, respiration and transpiration. Consider, for instance, the immune system, with its millions of different antibodies arrayed in defense of the body against millions of different intruders. This army must solve the fundamental problem of recognition: telling one's self (and one's friends) from everything else. And the problem has been solved in much the way human nations, and their armies, have solved the counterpart problem: by standardized, mechanized identification routines—the passports and customs officers in miniature are molecular shapes and shape detectors. It is important to recognize that this army of antibodies has no generals, no GHQ with a battle plan, or even a description of the enemy: the antibodies represent their enemies only in the way a million locks represent the keys that open them.

With this fanciful but plausible description in mind, imagine taking a dose of a vaccine against *Herpes zoster*—the chickenpox virus—and injecting it into a child who has not yet had the disease. The vaccine, which is a small and harmless portion of the virus, is immediately intercepted by lymphocytes—small cells that belong to the immune system, and we can consider them police patrols—intent on identifying everything that enters our organism from the outside.

The patrol does not know that the vaccine is harmless so the lymphocytes examine it carefully, as they do with any other unknown and suspicious substance. They then make a sort of three-dimensional digital cast of the vaccine, for future reference. Antibodies will, in a certain sense, be molded onto the cast down to the finest detail, just as a 3-D printer does with very small objects. From that point on, these molecules will be able to recognize other copies of the same vaccine, or very similar objects, and neutralize them. These antibodies will, for instance, go into action in the event of an attack by the real varicella virus, the same one from which the vaccine was derived and which it closely resembles, save for the fact that in its active and unadulterated form, it is far from harmless.

But what happens instead when *Herpes zoster* infects an individual who has not been vaccinated? The meeting between the virus and the lymphocytes occurs in approximately the same way as in the case of the vaccine, up to the 3-D cast phase. However, unlike the vaccine, which is harmless, the virus has a reproductive machine that in a matter of hours can produce millions of identical viral particles that then spread throughout the body.

The immune system's immediate reaction to the invasion is *inflammation*, a protective response whose goal is eliminating the intruder and initiating the healing process. Without inflammation no healing is possible. However, an uncontrolled inflammatory state can cause other problems. But that is not all, because the infection can also result in temporary or even permanent damage to the organs it has invaded, as well as open up the way for other pathogens. By taking advantage of an immune system that is already embattled, opportunistic microbes can find a favorable terrain for new infections.

## A vaccine is a bit like a flight simulator

Vaccination helps avoid all this by teaching the immune system to recognize a pathogen it has never encountered before, without the system suffering damage. When the virus, bacterium, or parasite against which the individual is vaccinated enters the body, the cells assigned to its defense remember the experience they had with the vaccine: they recognize the pathogen, neutralize it, and in this way completely avoid infection and disease. Thus, a vaccine allows us to acquire a harmless experience of the disease, creating a memory that will then be used in any future encounter with the pathogen, without the body having to suffer unpleasant, debilitating, or even lethal effects.

To better understand how the immune system learns from a vaccine, we can take the example of a pilot familiarizing with an airplane by practicing at the console of a flight simulator. Just as the inexperienced pilot avoids putting passengers at risk, similarly the vaccine allows the immune system to experience everything that can happen during an illness, without the body having to suffer its negative consequences. Essentially, the vaccine carries within it the necessary instructions for identifying the pathogen as well as the ways in which it can be fought in the same way as a flight simulator contains a broad sample of the problems and scenarios that a pilot could encounter during a flight.

In this chapter we have briefly observed the devastating damage that infectious diseases can inflict on populations, on civilizations, and on thriving economies. We have seen that pathogens can sometimes produce lightning blitzes while at other times they act more slowly, but with no less lethal effects.

These threats are counteracted by the powerful defense arsenal that humans possess against bacteria, viruses, and parasites. In this arsenal, vaccines represent a particularly ingenious weapon, capable on the one hand of exploiting the immune system's natural ability to recognize and memorize anything that is extraneous, while at the same time, sparing us from the unpleasant effects of diseases.

During an infectious disease, or when a vaccine encounters the cells of our immune system, the crucial phases of a very lengthy evolutionary history are reproduced within an accelerated timeframe: the same phases which, over thousands of years, pathogenic microorganisms and their imaginative camouflage strategies have had to face when dealing with the host's sophisticated mechanisms of recognition and defense.

# 2
# The Long History of Vaccines

Whenever a germ plays cops and robbers with our organism, the encounter involves complications, diversions, and twists worthy of an Agatha Christie novel. The defense system that protects us from disease has evolved over the course of a few million years, in parallel with the cunning tricks developed by microorganisms that have so often escaped our bodies' sophisticated surveillance systems.

If we are normally able to survive the ocean of germs in which we are immersed, it is largely thanks to several features of our immune system, such as physical barriers and molecular sentinels; inbuilt protections and defenses that update themselves based on their encounters; communication systems and postal services; on-call defenses and virus investigators; security systems and memorizers of molecular shapes. At the end of this book, you will find an appendix dedicated to our defenses and how they work.

Over time, we have incremented our natural abilities to prevent the deadliest infectious diseases, with the solutions and inventions that we will see in this chapter. These solutions were based on countless hypotheses formulated by ancient and modern scientists. For these hypotheses, experiments were made and repeated several times. For each experiment, a response: yes, this hypothesis is confirmed; no, this hypothesis is wrong; this hypothesis is perhaps confirmed, but to be sure, let us test it one more time.

This process, which is based on the so-called *scientific method*, allows us to explore the nature of life in a rigorous manner. The scientific method is a way of solving problems that each of us adopts every day, even without being scientists. For instance, if today the phone is not working, we begin formulating a series of hypotheses (the battery is low; the provider is having technical issues;

I forgot to pay the bill ...) that we test separately until we find a satisfying explanation.

That is exactly how the scientific method operates: from a complex problem, one variable at a time is isolated and tested. Each new piece of evidence, added and integrated with the others, slowly brings to light the overall picture defining the issue, just as the image of a mosaic emerges from the sum of its tiles. But unlike the mosaic, the overall picture of any complex scientific question can never be considered finished. The results of each new experiment can, in fact, lead to updates, adjustments, and corrections that change, improve, and clarify ideas and assumptions on a given topic.

In the past, physicians and scientists had adopted a sort of basic scientific method to combat infections, essentially made up of observations, deductions, and empirical evidence. Although primitive, it has been refined over time and has been surprisingly effective. We still have traces of the most archaic of these attempts from our remote past, a time when humans attributed the cause of diseases to noxious air, to witches, to the alignment of the planets, and other bizarre phenomena.

Louis Pasteur's discovery of germs as the cause of infection only took place at the end of the nineteenth century. But for the ten thousand years or so that came beforehand, medicine had not just stood by and watched. Countless remedies, antidotes, and therapies had become part of medical practice, long before we understood how and why they worked. The history of vaccines is no exception: methods for disease prevention were developed by trial and error, long before we were able to explain them fully.

## For ancient ills, ancient remedies: Variolation

"An illness never strikes twice: or, at least, a relapse is not lethal." With these words, the ancient Greek historian Thucydides observed how, during the epidemic that spread in Athens in 430 BCE, those who fell sick and survived would subsequently be immune to the disease. We do not know if the epidemic was caused by the plague, smallpox, or typhoid fever. Whatever the disease,

Thucydides noticed how survivors would tend to those who fell ill after them, as they were able to avoid a relapse of the disease.

The ancient Greeks had, in this way, witnessed the phenomenon of *acquired immunity*, although they were unaware of its causes: memory cells, still alive after a first encounter with an invader, ready to reactivate and proliferate as soon as that same microorganism reappears.

This fact was known in many other parts of the world besides Greece, especially in connection with a very particular disease: smallpox. The name will mean little or nothing to younger people, because as of 1979, when the WHO declared its disappearance, the disease no longer exists on Earth.

If you are between forty and sixty-five, you might have a small round mark on your shoulder, the "stamp" left by the smallpox vaccination that was obligatory almost everywhere between the late 1950s and early 1980s.

If you are in your seventies or older, the word smallpox probably evokes fewer reassuring memories for you. You have almost certainly known people who contracted the disease, unless you were unlucky enough to get it yourself, but lucky enough to recover.

The smallpox virus, with its two variants *Variola major* and *Variola minor*, probably first appeared in Africa around 10,000 BCE, at a time when our ancestors were beginning to domesticate animals. The unprecedented proximity with other species could have provided an opportunity for the first human contagion with a virus from this family, which includes bovine pox, camel pox, and monkey pox (despite its name, monkey pox mainly affects rodents).

By the seventeenth century, smallpox was endemic in Europe and Asia and most people had been exposed to the disease at some point in their life (the term *endemic* means that an infection maintains a constant presence at a basic level within a population of a specific geographical area, without being introduced from the outside). For every 100 people who fell ill with smallpox, about 30 died, but for children under 5, the death rate was even higher, over 90 percent in some regions. Joshua Loomis, a professor of biology at the University of East Stroudsburg, Pennsylvania, wrote that the danger was so great that "parents would commonly wait to name their children until after they had survived smallpox." For the eighteenth century alone, it is estimated that the disease caused the death of about 75 million people, around 10 percent of the world population at the time. Even in the last century, over 300 million people are estimated to have died of smallpox, when the infection was still as prevalent as cancer and cardiovascular diseases are today. But unlike these illnesses, smallpox killed far more children.

The virus was transmitted from person to person via droplets during close contact with people who were infected and had symptoms of the disease, or in some cases through contaminated clothing and bedding. Those who got infected would initially succumb to a fever before finding themselves covered with sores that would then turn into small, pus-filled blisters. If they did manage to survive, after immense pain, they would bear the visible signs of the disease on their bodies for life: the characteristic pockmarked face for almost everyone, blindness, and deformed limbs in the more severe cases.

Smallpox made no class distinctions: many monarchs and more than one pharaoh were affected by the disease. Facial pockmarks are still visible on the mummy of Ramses V (Figure 1). Mozart, Beethoven, George Washington, and

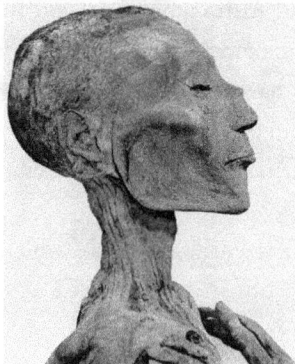

Figure 1: The mummy of the pharaoh Ramses V, on display at the Egyptian Museum of Cairo, still shows facial pockmarks (Wikipedia).

Abraham Lincoln survived the disease; Joseph Stalin, years before the advent of plastic surgery, hid his scars through regular photographic retouching. The microbes that cause infectious diseases are rather more anarchic than democratic: they strike at random.

The scourge of smallpox has accompanied the entire history of mankind; its rather ancient name is derived from the Latin *varius*, "spotted" or *varus*, "mark on the skin." Smallpox epidemics came and went in close and regular succession, and the threat of infection was an ever-present worry in the minds of men and women. On the one hand, this long period of cohabitation has terrorized families, cities, and nations with every contagion. On the other hand, human beings have come to know the disease pretty well, and over time have been able to observe certain characteristics that have proven important for prevention. Of these, the most significant was the fact that those who recovered from the infection became immune; that is, they did not get sick a second time.

Perhaps it was this very awareness that enabled the ancient inhabitants of China and India to develop a practice of smallpox prevention called *variolation*. People were intentionally infected in the hope this would cause a milder form of the infection, as opposed to the one contracted naturally, and that it would confer

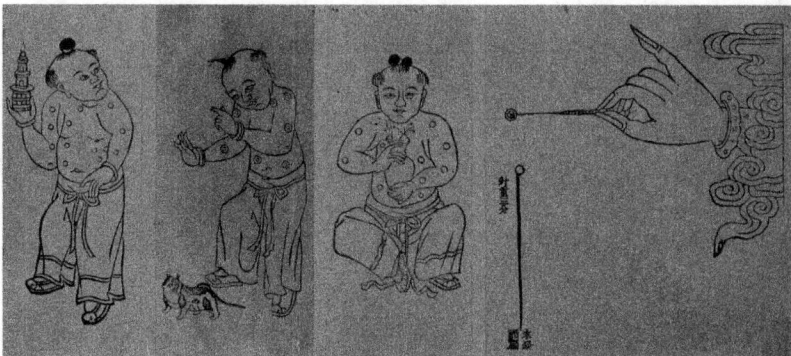

FIGURES SHOWING VACCINATION PUSTULES
From a Chinese work on Vaccination

Figure 2: An ancient Chinese illustration of variolation (Historical Medical Library of The College of Physicians of Philadelphia).

immunity. In this way, the imperial physicians of the Zhou dynasty, which ruled in China in around 590 BCE, would try to protect members of the court with this procedure: they would scratch dry scabs from an infected person, grinding them down to a fine powder which they would then blow into the nostrils of healthy people with a bamboo straw (Figure 2).

## Lady Montagu and smallpox in the Old Continent

By 1700, variolation was a common practice in Africa, India, and the Ottoman Empire, but it had not yet reached Europe.

Lady Mary Wortley Montagu was the young wife of the British ambassador to Constantinople when, in 1717, in a letter to an English friend, she described the variolation method in use at the time at the Ottoman court. The procedure described by the ambassador's wife was a development of the Chinese practice. Some elderly women working at the court would make four or five scratches on the arm of the person to be immunized, introducing the material taken from the pustules of someone suffering from a mild form of smallpox.

Lady Mary (Figure 3) was herself a survivor of smallpox. She had contracted the disease in England a few years before leaving for the East and had been left with permanent scarring on her face, in addition to mourning her twenty-year-old brother, who had died of the same illness. Perhaps because she had been personally affected by smallpox, Lady Montagu was determined to protect her children from the disease and to spread the practice of variolation back home.

And so, on her return to London in 1721, Lady Montagu was able to convince Dr. Charles Maitland, a doctor who had previously been at the Embassy, to inoculate her four-year-old daughter with smallpox-infected material, using the method practiced in Turkey. To *inoculate* means to introduce an infectious agent or an antigen into a living organism by injection or skin abrasion to produce immunity.

Dr. Maitland variolated the young girl with three physicians from the Royal College of Physicians in attendance, as witnesses. He then received a royal

Figure 3: S. Hollyer by J. B. Wandesforde, Mary Wortley Montagu in Turkish clothes (Bildarchiv Austria).

license to inoculate smallpox in six inmates of a London prison. They would act as "human guinea pigs" and in return, were promised pardon and release from prison. The experiment succeeded: the convicts survived the inoculation, indeed they failed to get sick after being exposed to smallpox and were subsequently pardoned and freed.

Despite the trial's "success," medical authorities remained skeptical and incredulous. They could not believe that progress in clinical knowledge could come from a culture that many considered quite primitive. Still, Maitland doggedly persisted.

One experiment was not enough. It remained to be seen whether the procedure could be safely repeated even among the youngest. And so, variolations were carried out on some orphaned children, again with success. As a result, Maitland was called upon to variolate the Prince of Wales's daughters.

In retrospect, we can now say that variolation, the first step of artificial immunization, was an unsafe practice. According to some estimates, up to 4 percent of people inoculated fell ill with smallpox as a result of the procedure, a risk that fails to meet modern-day safety requirements. Also, people who fell ill as a result of variolation could spread the disease, causing an epidemic. For these reasons, while the practice of variolation became known in most European countries as well as in North America by the end of the eighteenth century, it remained a little-used practice for quite some time.

## Cows and milkmaids

Around 1762, in a little country town near Bristol, a young doctor by the name of Edward Jenner had just completed his apprenticeship. One day Jenner overheard a milkmaid say: "I'll never find myself with a horrible, pockmarked face, since I've already had cowpox."

Cowpox was a benign disease for people, contracted by touching the pustules on the udder of an infected cow. The infection, which in humans would cause only a few red blisters on the body, would leave a precious legacy: the protection against smallpox.

In Jenner's time, however, this widely popular belief was discarded by doctors, despite being a widespread practice not only across the English countryside, but also in Italy, Germany, France, Holland, and Mexico.

Jenner, however, was intrigued and decided to do an experiment. On May 14, 1796, Jenner inoculated James Phipps, an eight-year-old boy, by making two incisions on his arm with the fluid taken from the cowpox pustules on a milkmaid. Sarah Nelmes, the milkmaid in question, had caught the disease from a cow named Blossom. In the weeks following the inoculation, Jenner exposed the child several times to smallpox, and the boy failed to get sick (Figure 4).

Figure 4: Ernest Board, Edward Jenner vaccinating James Phipps (ca. 1910, Wellcome Trust).

Today, intentionally exposing a child, inmates, or orphans, to the risk of a fatal disease would immediately land a doctor in prison. In Maitland's and Jenner's days, ethics and laws were less strict.

It is thanks to these first experiments on the efficacy of proto-vaccines that the popular beliefs assumed scientific value: infection with cowpox protected people against smallpox. Variolation with smallpox, considered unsafe, gave way to this new procedure with the less risky cowpox.

## The causes of infectious diseases

Jenner's method, which worked against smallpox, was, however, powerless against the many other infectious diseases that raged at the time. A major problem was that the actual cause of all infections, smallpox included, would remain unknown until the second half of the nineteenth century, notwithstanding the results of experiments and findings of some illuminated scientists, from the seventeenth century onwards.

In 1683, Anton van Leeuwenhoek, a Dutch scientist, had examined some dental plaque with a primitive microscope of his own invention. An entirely new realm of living entities, tiny and invisible to the naked eyes, had been discovered. But seeing is not enough: how do we know that the germs observable under a magnifying lens can be infectious agents?

Francesco Redi was personal physician to the Grand Duke of Tuscany, in central Italy, in the second half of the seventeenth century, when he conducted some tests with rotten meat. By putting the meat in containers, some closed, others open, he was able to observe how larvae appeared only on the meat in the open containers, where the flies had been able to deposit their eggs.

About a century later, Lazzaro Spallanzani, a scientist and a Jesuit at the University of Pavia, in northern Italy, was sterilizing infusions by boiling them for more than an hour. Some of the infusions were kept in flame-sealed glass containers, where Spallanzani noticed that bacteria did not grow.

Louis Pasteur was already a rather famous chemist in France when, in 1859, he conducted an ingenious experiment. He wanted to understand once and for all where the contamination of liquids used in food originated from. The hypothesis, still in vogue at the time, was that contamination was created out of nothing, according to the "theory of spontaneous generation," which dated back to Aristotle.

Pasteur filled a flask with some broth obtained from boiled meat, then heated the thin neck of the flask so that it curved like a swan's neck. In this way, the broth was still exposed to the air, but the flask's airway was both very narrow and tortuous. After three days, the broth was still uncontaminated: no germs had managed to travel along the twisted neck into the tempting broth.

In this way, Pasteur (Figure 5) demonstrated that contaminations do not arise out of nowhere: they only grow where microbes firstly manage to arrive, and then proliferate. The thousand-year-old theory of spontaneous generation ended with Pasteur's experiment, which followed the pioneering attempts made by Redi and Spallanzani.

Figure 5: Albert Edelfelt, Louis Pasteur (1885, Musée d'Orsay).

What does all this have to do with infectious diseases? Just as contaminants must enter into the broth—Pasteur reasoned—the germs that cause infections must penetrate into the host's body.

The echo of Pasteur's studies reverberated throughout Europe. At the University of Glasgow, in the 1860s, the surgeon Joseph Lister was concerned by the high death rate among patients in his ward. Lister hypothesized that the deaths were caused by infections from microorganisms carried from patient to patient by the doctors' unclean hands and from surgical instruments. After introducing the practice of washing hands and instruments with a disinfectant before surgery in his ward, the death rate dropped from 50 to 15 percent.

Meanwhile, in Germany, Robert Koch completed Pasteur's work by identifying the bacillus responsible for anthrax and tuberculosis (TB). This discovery earned him the Nobel Prize in Physiology or Medicine in 1905. Even more importantly, Koch devised a list of four criteria, or postulates, that must be met to determine whether a specific disease is caused by a particular infectious agent. The criteria are:

1. The microorganism must be regularly associated with the disease and its characteristic lesions.

2. The microorganism must be isolated from the diseased host and grown in culture.

3. The disease must appear when a pure culture of the organism is introduced into a healthy, susceptible host.

4. The same organism must be reisolated from the experimentally infected host.

Thanks to these criteria, every infectious disease can be attributed to a certain and specific cause: a pathogenic microorganism that can be isolated and identified. These postulates, used for a long time, are still very useful, even though they have some limitations that Koch himself identified right away.

## The legacy of Robert Koch

With cholera, Koch (Figure 6) was unable to meet all the requirements of his postulates, since *Vibrio cholerae*, the bacterial agent that causes the disease, presented itself in both sick and healthy people.

Even more complex was the case of diseases caused by viruses, which in Koch's time had not yet been discovered. Several viruses, like many bacteria, do not cause disease in all infected individuals. For instance, the poliovirus causes polio, but it paralyzes only 1 percent of infected people (today the disease is almost eradicated thanks to vaccines).

Also, while a single virus can give rise to several different diseases, many diseases present with the same symptoms despite being caused by different microorganisms.

Figure 6: Robert Koch (Wellcome Images).

Furthermore, the study of pathogens is not an easy task. Viruses are obligate parasites that need living cells to reproduce, but are often reluctant to grow in a cell culture. For other microbes, finding a laboratory animal that can be infected and manifest a disease sufficiently similar to how it appears in humans has yet to happen.

In short, a microbe that can satisfy Koch's postulates is probably the cause of the disease associated with it. But one that does not satisfy them cannot be excluded with certainty. The value of Koch's postulates, in other words, stems not so much from its standard and rigid application, but from the rigorous scientific reasoning required for collecting evidence and observations, even at the risk of violating the criteria themselves. This is why the criteria underwent an initial revision in the 1930s by Thomas Rivers, at Rockefeller University in New York.

Furthermore, there was a need for a second revision in the late 1990s, when it became clear that Koch's postulates were even more difficult to apply in the genomic era. Today, a machine can make multiple copies of genetic material from a tiny sample obtained in a diagnostic test. Other machines are able to rapidly analyze the sequences and tell us in a matter of seconds which piece of microbial DNA or RNA is present in the sample. We saw this extensively during the COVID-19 pandemic, when the name of these machines, reverse transcription polymerase chain reactions (RT-PCRs), resonated in the daily news. With these new technologies and testing capabilities we are now able to detect microbial genetic material even before the microorganism to which it belongs has been observed or isolated, or before its cultivation, whether bacterium or virus, in the lab. It is a bit like smelling sweet-scented banana in a smoothie, without ever having seen that long, yellow, and slightly crooked fruit from which the smell emanated, or without being able to grow the banana tree in an orchard.

Over the last 25–30 years, genomic feats have shown us that each human being is a hotel for microbes and that many bacteria and viruses, in both sick and healthy individuals, are still unidentified (on the genomic revolution applied to vaccinology, see Chapter 4). This increased diagnostic power, which has partially eliminated the need to grow microorganisms in a lab, is transforming modern medicine and microbiology. However, finding the genetic fingerprints of a microbe is not the same as proving a link of cause and effect between a microbe and a disease. In fact, this second step has possibly become even more difficult.

Take the hepatitis C virus (HCV) and the human papillomavirus (HPV), both recent success stories in terms of treatment and prevention. Today we know that HCV causes a form of hepatitis, and that HPV is at the origin of cervical cancer, but for years it has been impossible to cultivate either virus in a lab. The existence of hepatitis C was hypothesized in 1970, when it was called "non-A, non-B," a double denial revealing how medicine was ignorant and powerless against a virus that appeared impossible to isolate. Effective drugs are available today against HCV, and excellent vaccines can now prevent the most oncogenic strains of HPV. And this is despite the tribulation both viruses have given to dogged and determined scientists.

Koch's postulates were updated again by the microbiologists David N. Fredericks and David A. Relman in 1996, to consider both the opportunities and limits of the genomic era. The list of requirements has grown from four to seven and the language has become more complex, reflecting our increased knowledge of this vast and diverse microbial world.

According to a previous estimate, each member of the approximately 50,000 species of vertebrates is capable of hosting an average of around 20 endogenous viruses. This suggests the existence of at least 1 million different viruses, more than 99.9 percent of which are unidentified. Many can jump from species to species and cause disease in humans, as well as to domestic and wild animals. And that is only the viruses.

Every time a microbe duplicates itself and its genome, some genetic error, or mutation, occurs. Sometimes the replacement of just one nucleotide is enough to radically change the microbe's ability to pass from one host to another, or to cause a milder or more severe disease. However, the symptoms of a disease very much depend on the genetic susceptibility of the host, their age, their nutritional and general conditions, as well as any previous exposure to similar agents.

Proving a causal link between a microbe and a disease is particularly difficult when the effects are far-off in time or space, for instance, when a bacterium releases a toxin that begins to act long after the bacterium is gone. In other cases, cofactors or concomitant infections are needed for the virus or bacterium to cause disease. Under certain conditions, the causal link can be demonstrated only once an effective vaccine or treatment shows that it is possible to prevent or cure the disease.

Koch's postulates and their revisions are still relevant, not so much for their literal value, as for their lesson of rigorous scientific reasoning. By admitting that his method was perfectible, Koch reminds us that exceptions should neither frighten nor frustrate us. On the contrary, they are unique opportunities for adapting our way of thinking to the endless variety that makes up the biological world. Microbes, shaped over millions of years of evolution, will always remain elusive and a step away from our best attempts to keep up with them. Still, despite our own limitations, we are at times able to win some battles with them.

## Attenuated germs

Pasteur's contribution to the elimination of infectious diseases did not end with his experiments with broth in a swan-neck flask. After isolating several bacilli responsible for human and animal illnesses, in 1879 the French microbiologist was wrestling with chicken cholera, a bacterial disease he was studying by trying to grow several colonies of it in his lab.

Every time cholera bacteria were inoculated into poultry in the lab, they would, without fail, result in disease and death of the fowls. Always. Except once, when a particular bacterial colony was used to inoculate the birds, who remained alive and well. The same chickens were then infected a second time with fresh bacteria, but again, they failed to get sick with cholera.

What had happened? It transpired that the bacteria from this colony had been neglected over the summer break. At first, they grew rapidly, but then, left without nourishment, they became very weak or, as microbiologists say, *attenuated*. Inoculated in the chickens, these attenuated bacteria were unable to cause the disease, but the immune system spotted them anyway. In this way, the birds were immunized against any subsequent infection.

The idea that a milder disease can provide immunity to a more virulent form is not new. We already saw it, initially with variolation and later, with Jenner's immunization experiment with cowpox. The novelty, in this case, was that the milder form of the disease did not come from nature: it was caused by bacteria whose virulence has been artificially impaired by human intervention. The possibility of creating *in vitro*, and not only *in vivo*[1] bacterial varieties with

---

[1] An *in vivo* experiment uses a whole organism, under conditions as similar as possible to those in nature, whereas an *in vitro* experiment uses only a part of an organism (for instance, some cells or some genetic material) under controlled conditions.

reduced virulence were a significant leap forward for the science of vaccines. Thanks to their ability to immunize against the more aggressive forms of diseases, these attenuated microbes greatly assisted the development of vaccines that came into use during the twentieth century.

Pasteur named the artificially attenuated bacteria *vaccines* in honor of Jenner's discovery. The term derives from the Latin *vacca*, meaning cow. Jenner had used the derivative, *vaccine*, to indicate both the cowpox that affected cows and the origin of the protective liquid against smallpox, which he had extracted from the pustules of sick animals.

Having coined the name, Pasteur began producing a large batch of vaccines that would be adopted extensively over the next century. First in the series: the anthrax vaccine.

## Rabid bites

Joseph Meister was a nine-year-old boy when, in 1885, he arrived in Pasteur's lab accompanied by his mother, after being bitten by a rabid dog. Rabies is a viral disease, usually transmitted to humans from the bite of an infected animal. At the time it was quite rare, but always lethal.

Emile Roux, a physician and colleague of Pasteur's, had a rabies vaccine in his icebox. He had developed it by infecting some rabbits with the virus and then extracting some virus-infected tissue from the animals' dried spinal column. Desiccation was another technique used in Pasteur's lab to transform pathogens into less virulent strains. In the case of the rabies vaccine, the viruses had been utterly destroyed by desiccation, but the ability to generate immunity was intact, as Roux had already demonstrated by testing the vaccine on dogs.

The rabies vaccine had yet to be tested on humans, until Joseph Meister showed up in the lab. The boy, left untreated, would have certainly died. After consulting with his colleagues, Pasteur decided to administer the vaccine. The attempt was a big gamble for Pasteur, who was not a doctor and risked being prosecuted if the experiment failed. Fortunately, the vaccine worked, and the immense resonance of the case paved the way for the use of vaccines among humans.

## Harmless toxins and pathogens

Diphtheria is a disease we rarely hear about today. Ten cases of diphtheria and one death were reported in England in 2019, according to statistics from Public Health England, but up until the mid-1950s, every year tens of thousands of people would fall ill with the disease, which was one of the leading causes of death for children under fourteen.

The bacillus that causes diphtheria, *Corynebacterium diphtheriæ*, is harmless in itself, but the powerful poison it releases in the host's body is not. This poison, known as the diphtheria toxin, mainly affects the heart and nervous system, causing serious harm, and in many cases, death. Even small amounts of the toxin are lethal (a dose of 0.1 micrograms per kg of body mass is enough to kill a human being).

The first therapy against the diphtheria toxin was the *therapeutic serum* or *antiserum* that Emil von Behring, a German physician and bacteriologist, developed at the end of the nineteenth century.

Serum is the liquid component of blood and it, among other things, contains antibodies. Behring obtained the antiserum by administering an attenuated form of the bacilli to laboratory animals. The antibodies that the animals developed in their serum were able to deactivate the diphtheria toxin, even when the serum taken from these animals was injected into other animals infected with the nonattenuated bacteria.

Behring's discovery, which earned him the Nobel Prize in Physiology or Medicine in 1901, has been the first in a long series of antisera-based therapies. But the immunity provided by antisera vaccines remains *passive*: it is the antibodies present in the serum being injected that are at work, not those developed by the person affected by diphtheria. In other words, in Behring's time there was still no vaccine against diphtheria. A vaccine that, when inoculated in a human being, would trigger the production of an *active* immunity on each encounter with the germ, made up of antibodies against the diphtheria toxin.

A huge obstacle lay in the way of such vaccine was that the toxin had to be inactivated and transformed into a harmless antigen. Harmless, but still capable of inducing the immune system to produce adequate antibodies against future infections.

Gaston Léon Ramon was a French veterinarian who, in 1924, found the solution to this problem: by subjecting the diphtheria toxin to a treatment with form-aldehyde—a chemical substance—at 37°C, the toxin became harmless, its immunizing power intact.

Using this same procedure, Ramon obtained a second vaccine against the toxin produced by *Clostridium tetani*, the bacterium that causes tetanus. Tetanus is a noncontagious disease contracted through a contaminated wound. The tetanus toxin causes the prolonged contraction of skeletal muscle fibers and, if left untreated, can lead to a terribly agonizing and painful death.

From these two applications, Ramon's method, which was reliable and repro-ducible, would go on to be widely adopted to inactivate many pathogens that have been developed into vaccines. It was another step forward, towards the mass vaccinations that characterized the twentieth century.

Ramon's two preparations also demonstrated another important principle for the future development of vaccines. They were the first examples of preparations that worked by inoculating a *subunit*, or a part, of the entire pathogen.

## Polio and mass vaccinations

Poliomyelitis, or polio, is a contagious and viral disease caused by the poliovirus, which is spread from one person to another through contact with feces or by mouth. The symptoms of poliomyelitis are motor paralysis—localized mainly

Figure 7: Representation of a victim of polio in an Egyptian stele of the Eighteenth Dynasty (Deutsches Grünes Kreuz).

in the legs—but can affect other parts of the body, depending on where the infection is spread in the central nervous system.

Egyptian stelae, depicting people with legs withered by paralysis and children walking with the help of a stick (Figure 7), tell us that the effects of polio have been known since the ancient times. However, for thousands of years, the disease was not considered as serious a threat as smallpox, the plague, or other common infections, because the virus epidemics were limited and sporadic.

The notion suddenly changed at the beginning of the twentieth century when serious epidemics erupted: first in Europe and then in the United States, reoccurring with increasing intensity until well beyond the middle of the century. At the same time, the age group of the people most affected also changed over the 50 years or so in which the epidemics intensified, initially hitting younger children hardest, but by 1950 targeting children between 5 and 9, an age in which, among other things, the risk of paralysis was greater. Even adults were not spared: about a third of cases involved people over 15.

The worst epidemic was in the United States in 1952, with about 58,000 cases, of which over 3,000 were fatal. Thousands of people around the world remained paralyzed for life.

For the first time in history, people's suffering and anguish achieved enormous visibility thanks to powerful new means of communication: radio and television. The widely broadcast emergency provided the impetus to develop a vaccine for administration to the entire population.

After decades of research and experiments, in 1952 the first vaccine emerged from the lab of Jonas Salk, and in 1955 it was launched through mass vaccination campaigns. The Salk vaccine contained a poliovirus grown *in vitro* in the cells of monkeys' kidneys and subsequently inactivated using the method developed by Ramon.

"Who owns the patent on this vaccine?" Jonas Salk was asked by a TV journalist right after the vaccine was produced. "Well, the people, I would say, there is no patent … Could you patent the Sun?" Salk had chosen not to profit personally from his discovery.

The second vaccine, produced in 1957 and approved in 1961, was developed by Albert Sabin (Figure 8), with a poliovirus that he attenuated by letting it multiply several times in nonhuman cells, at temperatures below those preferred by the pathogen.

## Towards eradication

The anguish experienced by people in many countries during the polio epidemics, until the 1950s, were the base of the strong support and demand for mass vaccination programs against infectious diseases. Pasteur's principle and Ramon's method can be summed up with the motto "isolate the germ, kill it,

Figure 8: Albert Sabin at the entrance to the virology department dedicated to him at the formerly Sclavo Institute in Siena, Italy, in 1962 (Sclavo Institute).

inject it." This principle allowed for the large-scale production of other vaccines in addition to those we have discussed so far.

Thanks to vaccines, smallpox had been completely eradicated from the face of the Earth within a century. The last known person to contract the disease was Ali Maow Maalin, a hospital cook in Somalia, in 1977. He would recover from smallpox, only to die of malaria in 2013.

Since the adoption of the antipolio vaccines, cases of wild poliovirus have decreased by more than 99 percent since 1988. At that date there were around 350,000 cases in more than 125 countries where the virus was endemic, while only 33 cases were reported in 2018. Of the three wild poliovirus strains (types 1, 2 and 3), type 2 was eradicated in 1999 and no cases of type 3 have been reported since the last one was identified in Nigeria in November 2012.

Thanks to vaccines, cases of diphtheria, measles, rubella, mumps, pertussis (whooping cough), diseases caused by *Hæmophilus influenzæ*, and tetanus have been reduced by more than 98 percent. To make a comparison, the most effective medicines are those used to treat rheumatic fever and rheumatic heart disease, with 75 percent effectiveness.

These impressive numbers were achieved after thousands of years of attempts, experiments, and clinical trials which received a powerful boost from the end of the nineteenth century onwards.

The heroes of this story are certainly the scientists involved, with their brilliant minds, and Lady Montagu, with her strong power of persuasion. But we must be equally grateful to all the inmates and the orphans, to James Phipps, to Joseph Meister, to Lady Montagu's own children, and to all the people who acted as human guinea pigs for these therapies, unaware of the risks they ran.

Today it would be impossible to perform similar experiments on humans, especially minors, with the same level of risk that those tests entailed. Popular sensibilities as to the value of human life have evolved over the last century, giving rise to laws that scrupulously regulate the testing of drugs, vaccines, and therapies on humans. But without those early contributions, many of us might not be here to write or read this book today, while the value of life itself would no doubt be perceived quite differently.

# 3
# Listen to This

B y the 1950s, vaccines seemed to be on the verge of defeating infectious diseases forever. The scientists who had developed them were the heroes of the day, appearing on the covers of the most popular magazines, with radio and TV programs competing to interview them (Figure 1). Vaccines were now saving more lives than any other medical intervention, turning some serious and deadly diseases into a past memory.

The public experienced a sort of honeymoon with vaccines, which they considered almost omnipotent. But vaccines, although precise and effective, are still human artifacts, and as such, are not infallible. Their benefits far outweigh the risks, but in very rare cases they can give serious and even fatal side effects. The Sabin vaccine, for instance, would reacquire its aggressive form in 1 out of 750,000 inoculated people, causing paralysis, because it contains an attenuated—as opposed to an inactivated—virus.

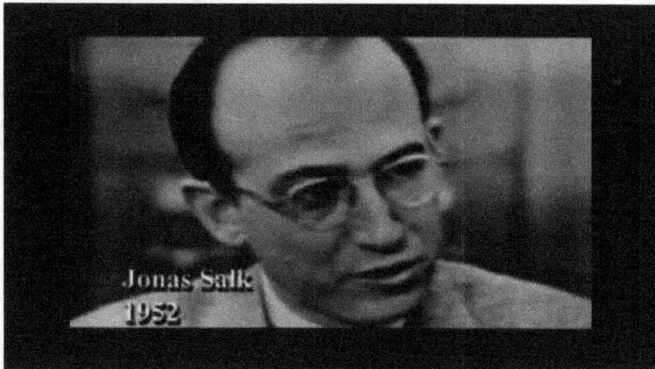

Figure 1: Jonas Salk as a guest of a television program in the 1950s (YouTube).

## Do we know how to evaluate the risks?

Is 1 in 750,000 a large, medium, small, or very small risk? Numbers alone are insufficient. To understand their significance, it is useful to compare them with some daily risks we are probably more familiar with.

Let us take, for instance, the probability of getting hurt in a car accident or at home: in the United Kingdom in 2018, 1 person in about 400 was involved in a car accident, while every year 1 out of 25 people visits the local emergency ward after suffering a domestic accident.

By comparison, the probability of becoming ill with polio after the Sabin vaccine is *almost two thousand times lower* than the probability of being in a car accident and *thirty thousand times lower* than the risk of needing to go to an emergency ward following a home injury.

Yet you will find that many people are more afraid of having their child inoculated with the Sabin vaccine than of driving the same child around in a car or having them play freely at home.

Sadly, the human brain does not excel in risk assessment. Numbers, perhaps because they are too abstract, do not seem to work for guiding our choices. Above all, they fail to convince us that the bite of a poisonous spider is a far lesser risk than that of being hit by a car. We are more afraid of a very rare but sensational risk, say a tsunami, than of a trivial but frequent danger, like slipping in the shower. And we are far more scared of a sudden event, no matter how unlikely—such as the potential side effect of a vaccine—than of something that progresses steadily but slowly over time (like the carcinogenic tar that builds up in the lungs of a smoker, or the gradual narrowing of our arterial lumens from too much fat in our diet).

Today, in many countries, including the United Kingdom, only the Salk vaccine is used to vaccinate children against polio, to minimize the already negligible risks associated with the Sabin vaccine. And this is despite the latter vaccine's many advantages: it is cheaper (a not insignificant fact when large populations have to be vaccinated); it can be administered by mouth rather than by injection (an important advantage in countries with poor health facilities and few medical personnel); and it provides a broader protection over time, especially in the intestine, which is the typical entry point of the poliovirus, a critical issue in countries—such as Afghanistan and Pakistan—where the virus is still endemic.

## Safety and surveillance

Before being authorized for use, a vaccine is effectively tested on a number of individuals, ranging from 5,000 to 100,000. However, rare side effects may not appear during clinical trials—for instance, if they appear in less than 1 in 5,000. For this reason, public health authorities in most countries are responsible for monitoring possible side effects from vaccinations that could emerge once the vaccines are in use among the general population, outside of the clinical trials.

In 1998, in the United States, a vaccine against rotavirus (a microorganism that kills more than half a million children every year in developing countries, causing fever, vomiting, and diarrhea) was at the origin of a severe intestinal blockage in about 100 of 1,000,000 inoculated children. Epidemiologists became aware of the problem within months of the vaccine's introduction. The Food and Drug Administration (FDA) immediately withdrew its authorization, and the US Centers for Disease Control and Prevention (CDC) removed the vaccine from the list of recommended shots.

The US CDC's Immunization Safety Office is responsible for research on vaccine safety, for clinical monitoring, and for providing timely and transparent information to public health officials, healthcare providers, and the public, on its website.

Overall, vaccines have a very low rate of serious adverse effects, but in such cases the response capacity is quick and effective. This is very good news, especially when one thinks of the hundreds of thousands of disabilities and deaths that would be caused by infectious diseases if we were not vaccinated, and therefore protected. Yet for many people, especially in developed countries where many of the diseases against which we vaccinate are now rare, the mere fact that a vaccine can involve even a minimal risk is unacceptable.

## (False) vaccine alert

In 1998, Andrew Wakefield, at that time a British gastroenterologist, published an article in *The Lancet* medical journal where, based on research on eight autistic children, he hypothesized that their condition could have been caused by the trivalent[1] vaccine against measles, mumps, and rubella (the so-called MMR vaccine).

The reaction to Wakefield's article was explosive: it had been published in *The Lancet*, one of the most authoritative and credible medical journals in the world; it was about autism, a disorder with very uncertain causes; it concerned the vaccinations which millions of children are subjected to every year. The findings were widely broadcast by newspapers, radio, and TV in almost every country in the world. The outcry was huge and caused much alarm among parents who were about to have their children vaccinated.

In 2000, another article, published this time in *Medical Hypothesis* (a far from authoritative journal), floated another disturbing theory: that thiomersal, a preservative derived from mercury and used in vaccines, could cause autism.

## What is autism?

First of all, let us start with a terminological clarification: the generic name of *autism* in fact covers a considerable number of related conditions, which vary in the scope and severity of their symptoms, to the point where today's scholars prefer to talk about *autism spectrum disorder*. The most severe form of these disorders is autism, while the mildest is referred to as Asperger's syndrome.

What are the symptoms of autistic disorders? While the symptoms are extremely variable, some are common to most of the conditions falling within the

---

[1] A vaccine is said to be polyvalent when it protects against more than one infectious agent; trivalent if it provides protection against three pathogens, and bivalent if it defends against two.

spectrum. The most frequent are: a continuous repetitive behavior of some gestures; a narrow focus of attention; and a difficulty in communicating and interacting with others.

Autism spectrum disorders typically present themselves within the first three years of life, at around the same time that children are vaccinated. Parents of autistic children sometimes seize on that coincidence and convince themselves of a relationship between the two events, but as we shall see, it is in fact only a temporal coincidence, and not a link between a cause and an effect.

## Mercury in vaccines: Why?

Someone, hearing about mercury, might believe that the very idea of introducing this substance into our body—for whatever reason—is distressing. Mercury is indeed a powerful neurotoxin, and poisonous to our nervous system. Its toxicity has been known since the 1950s, when a chemical factory discharged massive doses of this substance into the sea off Minamata Bay, in Japan. Over three thousand people who fed on fish and other marine animals caught in the bay became seriously ill and about six hundred died. The cause—it was later discovered—was mercury poisoning.

Given that mercury is a neurotoxin, why was it used as a vaccine preservative? The decision goes back some fifteen years prior to the Minamata accident. Today, vaccines are almost always packaged in single-dose vials to be used for a single vaccination, but for much of the last century, industries had produced multiuse vials for more than one injection. Doctors and nurses extracted each dose from the vial using a syringe, with the needle entering the vial through a rubber stopper. The syringes were not disposable and so more than once it happened that during this process, the vaccine would become contaminated with bacteria.

Following two serious episodes of infection in the United States, in 1916, and in Australia, in 1928, health authorities in many countries established that vaccine multidose vials must contain a substance capable of killing any bacteria.

For a while, the choice of antibacterial substances remained rather limited (antibiotics did not yet exist), and at the time some mercury derivatives were among the most effective antiseptics (some among you will perhaps remember mercurochrome, a bright red wound disinfectant). For vaccines, the choice fell on ethylmercury, an effective and nontoxic preservative, combined with another substance (thiosalicylate), to form what was then called thiomersal.

How do we know that thiomersal is nontoxic? Firstly, the substance was tested in animals before being administered to humans. Also, thiomersal had already been used in humans in 1929, during a meningitis epidemic, in the hope that it could contain the infection. It was then discovered that, although thiomersal is unable to kill the bacteria that cause meningitis, it is in fact safe: people injected during that epidemic with a dose of two million micrograms (ten thousand times the dose contained in vaccines) showed no signs of mercury poisoning.

If the ethylmercury contained in thiomersal is not a neurotoxin, why was the mercury dissolved in the waters of Minamata toxic? The chemical formula of *methylmercury*, the substance discharged into the sea by the factory, is $CH_3Hg^+$, while for ethylmercury, it is $C_2H_5Hg^+$: the two molecules differ by one carbon and two hydrogen atoms. This difference, while minimal, is what makes methylmercury harmful.

Very small chemical differences can be critical with other kinds of molecules too. For instance, in Italy, during the 1980s, several people died after drinking wine that had been adulterated with methanol by its producer. Ethanol, the alcohol that is normally found in wine, differs by only a $-CH_2$ group from methanol, but it is precisely that group of three atoms that makes methanol a powerful poison.

Coming back to the different forms of mercury, ethylmercury is not toxic because it is quickly eliminated from our body, while every time we eat methylmercury, with contaminated fish for instance, it accumulates in our body, as we are unable to dispose of it as effectively.

## Studies on vaccines, mercury, and autism

Does a child who has been vaccinated with the MMR vaccine run a greater risk of becoming autistic as opposed to a child who has not been vaccinated? And is a child more likely to get autism spectrum disorder if they have been injected with a vaccine preserved with thiomersal, compared to a child whose vaccine did not contain that substance?

To answer these questions, since the mid-1990s public health authorities in many countries have been carrying out rigorous and controlled studies on millions of cases of vaccinated and unvaccinated children. In one of the most recent studies, a group of epidemiologists led by Dr. Anders Hviid, of the Serum Institut in Copenhagen, observed 657,461 children born in Denmark between 1999 and 2010, 31,619 of whom were not vaccinated. Over time, 6,517 of these children received a diagnosis of autism, but the incidence of this condition was no higher in vaccinated children than in those that were unvaccinated. These results, published in 2019 in the *Annals of Internal Medicine*, confirm an earlier study by the same group, regarding 537,303 children and published in 2002 in *The New England Journal of Medicine*, the most authoritative of all medical journals.

In addition to the Danish reports, more than 40 studies have been conducted in several countries, including the United States, Canada, Finland, and Sweden, to name but a few. In all these studies, the answers to both questions have been always negative and highly reassuring: the number of cases of autism spectrum disorder was the same for children vaccinated against MMR and for children who were unvaccinated. Analogous results were observed among children who had or had not received vaccines preserved with thiomersal. In short, there is no evidence that vaccine administration increases the risk of autism or that it triggers autism spectrum disorder in children with a predisposition to the condition.

Both scientists and public health officials considered the results of the studies, which in the meantime had been published in many leading journals, to be so convincing as to speak for themselves. Vaccines do not cause autism: no link has been found between the side effects of a vaccination (such as temporary fever) and autism spectrum syndromes; the thiomersal preservative that was used in the multi-dose vials for pediatric vaccines is not at the origin of autism spectrum disorders; and autism rates have continued to rise, despite the fact that thiomersal was removed from all childhood vaccines in 2001 (purely as a precautionary measure).

Yet when epidemiological studies of the MMR vaccine and thiomersal failed to show any association with autism, alternative theories emerged. The most popular theory, again proposed by Andrew Wakefield, suggested that the immune system would be overwhelmed and weakened by a simultaneous administration of multiple vaccines, creating an interaction with the nervous system capable of unleashing one of the autism spectrum disorders in a susceptible host.

## Are multiple vaccines given together safe?

Combinations of multiple vaccines induce the same immune responses as those resulting from the same vaccines when administered singly. And a young child's immune system, although relatively immature, is already capable of generating a vast array of protective responses.

Additionally, several technological advances have now made it possible to reduce a vaccine's *immunological load*. The current vaccines that are administered routinely to children in developed countries contain a total of less than two hundred viral and bacterial proteins[2] or sugars. For the sake of comparison, in the 1980s these components were more than three thousand in half the number of vaccines.

Why has the number of components in each vaccine been reduced by so much? Well, you could say that twentieth century vaccines were rather crude preparations: whole microorganisms, killed or attenuated, were injected, because at

---

[2] Proteins are large biological molecules, made up of chains of building blocks, called amino acids, whose sequence is determined by a specific gene. They are essential for the structure and functions of living organisms.

the time it was not yet possible to select the most antigenic parts of a virus or bacterium. Today, this is no longer the case: vaccines are designed using computer models and laboratory experiments that allow for the insertion of only those parts necessary to stimulate an immune response.

The total number of components found in routine vaccinations is also tiny compared to what a child's immune system has normally encountered by the age of five. Over this period, a child falls sick on average with four to six infectious agents per year, each containing an average of two thousand proteins. The immune responses that children unleash against this wide and varied exposure to wild infectious agents far exceed those stimulated by vaccinations.

Ultimately, multiple vaccinations do not weaken the immune system, since vaccinated and unvaccinated children are equally susceptible to infections that are not prevented by vaccines. In short, we have no evidence to date showing that vaccines can weaken or overwhelm our defenses.

But insufficient evidence is not the same as an acquittal, and at least a part of the public remains unconvinced.

## Politicians, actors, and media against vaccines

Since the 1990s, the same television talk shows that had been competing for an interview with Salk or Sabin forty years earlier now began hosting people claiming that vaccines were weapons of mass destruction. According to a theory advocated by activists such as Robert J. Kennedy Jr., pharmaceutical companies were in league with public health authorities to ensure that children around the world contracted autism.

Conspiracy theories are great for TV ratings, with stars like Oprah Winfrey and Larry King having ignored the most authoritative scientific evidence that exonerates vaccines.

These rumors have recently gained additional backing from well-known political figures in search of a consensus, including Donald Trump. Even a great actor like Robert De Niro has supported the hypothesis discredited by research, a sign that vaccines have reached a peak of unpopularity. A great rift has arisen between the science that exonerates vaccines and public opinion, with celebrities and media, hungry for visibility, adding fuel to the fire.

The irrational fear that vaccines can produce abnormal, undesirable effects has ancient roots, as illustrated by this satirical illustration from the early 1800s (Figure 2). Today, the myth that links vaccinations to autism and other imaginary risks is continually amplified on websites and social networks, giving new life to an ideology able to find followers and frighten confused parents. It has even been alleged that some countries' secret services have exploited these baseless ideas, helping to spread them through the use of bots and other powerful technologies on the web, with a view to foment a climate of confusion, destabilization, and chaos in nations perceived as enemies.

*The Cow-Pock — or — the Wonderful Effects of the New Inoculation! — vide. the Publications of ye Anti-Vaccine Society*

Figure 2: James Gillray, Collective hysteria for the smallpox vaccine (1802). The image makes fun of the fanciful and baseless idea that the vaccine could turn people into cows (Library of Congress).

## The search for causes and treatments of autism spectrum disorders

Let us be clear: the desire for parents of autistic children to find causes and cures for their child's condition is legitimate and understandable.

It is estimated that 1 in every 160 children worldwide is affected by an autistic disorder. This estimate represents an average figure for the numbers reported by various studies. However, incidence shows strong geographical variations, and data are missing for poor countries, for lack of research. Despite these gaps, the prevalence of autistic syndromes appears to be increasing globally, wherever epidemiological studies have been conducted over the past 50 years.

The reasons for the increase are unclear, but they seem to be, in part, due to changes in diagnostic criteria. Today, the umbrella of autism spectrum disorders includes severe forms, previously diagnosed as psychosis, and forms with mild symptoms, once considered personality quirks. This could go some way towards explaining the increasing number of cases.

Over the past twenty years, many families affected by autism have set up discussion and action groups for mutual support and to look for causes and treatments for their children's condition.

Within these groups, understandably sensitive to any theory as to the causes of autism and possible therapies, theories linking vaccines to autism have gained a lot of attention.

Less understandable is the continuing pressure exerted by some of these groups against the use of vaccines, despite the lack of any evidence for the alleged role of vaccines from any of the serious scientific studies conducted to date.

However, not all these groups sing the same tune. For example, the Autismo Treviso Association is an Italian patient organization for the families of autistic children that included, as one of its founders, Fabio Brotto, father of Guido, a young boy with a serious form of autism. The scientific information on the Association's website and Facebook page is reliable and the many theories spread by charlatans on the relationship between vaccines and autism are refuted with simplicity and effectiveness. An effort undertaken with the aim of protecting the community of families affected by these conditions against dangerous and deadly misinformation. Sadly, Fabio Brotto died of COVID-19 in 2020.

## Mercury is taken out of vaccines!

Under highly charged public pressure, in 2001, thiomersal was removed from pediatric vaccines as a precautionary measure in the United States, despite the lack of any evidence that thiomersal caused any risk or danger.

"If you remove cars from the highways, you will see a marked decrease in auto-related deaths," Robert Davis commented in 2008, when he was head of the US CDC's Immunization Safety Office. "If thiomersal were a strong driver of autism rates and you remove it from vaccines, [researchers] would have seen some sort of decline—and they didn't."

Yet even this additional proof held little sway over people convinced that vaccines were responsible. What else lay behind this hardened misconception?

## Fraud and manipulation

In 2004, some disturbing facts began to emerge. Andrew Wakefield and other supporters of the alleged relationship between vaccines and autism had received large sums of money from law firms specializing in class actions. A close collaborator of Wakefield testified that the experimental data on which *The Lancet* article was based had been artfully manipulated: the article was a fraud. Following this disclosure, Andrew Wakefield was fired by the London hospital where he worked and was banned from practicing medicine by the British Medical Council.

The emerging picture was disturbing: for years, lawyers, insurance appraisers, public relations experts, and unscrupulous charlatans had been spreading fraudulent "scientific" evidence in an effort to persuade parents of autistic children to initiate class actions against pharmaceutical companies producing the "incriminated" vaccines and the public health authorities recommending them. They hoped to obtain billions in compensation, exploiting the tragedy of families with autistic children, the hypothetical victims of vaccines.

You would think the story would end here, but unfortunately this was not to be, thanks also to media's failure to report on these frauds with the same emphasis used earlier to publicize the theory that vaccines caused autism spectrum disorders.

## On the run from vaccinations

Now that you know the story and have a better idea of how it actually went, try typing the word "vaccines" into Google or any social media. You will find multiple pages and sites that still today report false and misleading information as to the alleged association between vaccines and autism. The damage caused by the persistence of this incorrect information is difficult to estimate, but it is far from trivial.

In the months following the release of Wakefield's *Lancet* article, the percentage of children vaccinated against measles in London had dropped by nearly 50 percent. Between 2006 and 2007, over 10,000 cases and more than 1 death from measles—a totally preventable disease—were registered in Germany, Switzerland, the United Kingdom, Italy, and Romania. In 2019, 1,092 cases of measles, 6,360 of mumps, 4,613 of whooping cough, 12 of diphtheria, 3 of rubella, and 4 of tetanus were reported in Great Britain and Northern Ireland alone. All of these diseases, and the suffering and disabilities they can cause, can be prevented with safe and reliable vaccinations, offered free of charge by the NHS.

This pulling back from vaccinations—the most effective public health tools we have, surpassing even the best medicines—is still ongoing today, almost 25 years after those reprehensible acts. According to an international survey conducted by the Ipsos MORI Institute in 2017, in the United Kingdom, 20 percent of the people interviewed believed that some vaccines cause autism, 35 percent did not know, and 45 percent were properly informed. The percentages in other countries are no more reassuring: in the United States, 19 percent of those who participated in the survey believed that vaccines cause autism, in Sweden 25 percent, in Israel 29 percent, and in India 44 percent, despite the overwhelming lack of scientific evidence.

In the meantime, the risks of epidemics from vaccine-preventable diseases are still high where children are not vaccinated. In the first six months of 2019, the United States reported the highest number of measles cases in 25 years, while almost 90,000 cases were reported in Europe, more than those registered in 2018.

In spite of this, we continue to devote a lot of time and public money to theories that have no way of being proven, because they have repeatedly been shown to be spurious. Those resources would be better used in serious research into the causes of autism. Studies on this unsolved problem have recently yielded important results. In 2018, the largest genetic analysis ever carried out on these conditions, conducted over more than 10,000 individuals and led by a research group at the Children's Hospital of Philadelphia, revealed that at least 90 different genes are likely candidates for a causal role in predisposing children to the development of autistic syndromes. Research has also shown other risk factors, including the advanced age of both parents, some problems during pregnancy, and certain birth complications. Taken together, the results of these studies further exonerate vaccines as potential causes of autism spectrum disorders.

## Are vaccines victims of their own success?

The consequences of mass vaccinations, at least in developed countries—alongside hygiene, medicines, and a varied and balanced diet—are zero infant mortality, fewer minor illnesses in the first years of life and an expectancy of good health up to a very old age. We have come to rely on these benefits without giving it a thought: we take them for granted, as if they had always been there.

It has been shown that the widespread use of mass vaccinations causes a population to lose awareness of the severity of the diseases, because their effectiveness is measured more by nonevents than by events: the polio infection that *did not* leave my child paralyzed or the meningococcus who *did not* kill my sister. In a way, we could say that vaccines are victims of their own success: they have eradicated many very serious infections so well that subsequent generations with no experience of these diseases have neither memory nor a fear of them.

Of those parents who fear vaccinations more than the diseases they cause, no more than 10 percent reject them categorically. Most are simply confused, hesitant, and misinformed: perhaps they have read something online or have heard of a serious reaction to a vaccine and are worried. Others have little faith in the government and its impositions or recommendations. Still, others refuse some, but not all, vaccines—or prefer for their children to be vaccinated with a delayed vaccination schedule. This is despite the fact that the official schedule has been specifically designed to offer immunizations when they are most needed by children to protect them from infectious diseases.

And so, when having to choose whether or not to have their children vaccinated, many parents in these cases decide not to. They think that the risk is not worth the benefit, having no direct experience of many vaccine-preventable diseases, which have all but disappeared in their country. We forget that vaccination is a tool for *controlling* viruses and bacteria: if we stop vaccinating ourselves, are we sure that the invaders will not quickly regain their reproductive vigor?

A tangible example of what can happen when vaccinations are suspended can be found in the recent and tragic events in Venezuela. There, a dictatorship has plunged the country into a humanitarian and economic crisis of exceptional proportion. No cases of measles or diphtheria had been reported in decades, but both diseases have reappeared following disruptions to the national immunization program, triggered by the collapse of the health system. From June 2017 to October 2018, 7,524 suspected cases of measles and 2,170 cases of diphtheria were recorded in Venezuela, the latter with a mortality rate of 22 percent, as vaccination coverage for both diseases fell by 50 percent.

Something similar had happened in the past, in the countries of the former Soviet Union. In Russia, diphtheria was all but forgotten, thanks to the universal vaccinations that had begun in the late 1950s, as in other industrialized countries. With the fall of the regime in 1989, and the chaos that ensued in the health system, mass vaccinations came to a halt. Against this setting, diphtheria toxin made a big comeback, with 140,000 cases and 4,000 deaths in the early 1990s. This was the first large-scale diphtheria epidemic in an industrialized country, after more than three decades of "silence" from the bacterium.

## The decline in herd immunity and the return to mandatory vaccination

Each avoidable contagion can provoke the danger of a new outbreak in the population, in addition to the risk of disease and death for the individual. The higher the number of people vaccinated within a population, the greater the protection that even non-immunized individuals indirectly receive. This effect is called *herd immunity*.

An example of herd immunity comes from the United Kingdom, where the vaccine against a form of meningitis caused by meningococcal C (Men C), a bacterium, has been available since the end of 1999. Not all British citizens are vaccinated against it, but as a substantial part of the population has been, even those who have not been vaccinated (for instance, infants under three months, people with certain illnesses or people who have chosen not to get vaccinated) benefit from the cover.

That said, it would be a serious mistake to forego vaccination with the idea of benefiting from those of others. Herd immunity only works if a sufficiently large number of people is vaccinated, but it stops spreading benefits if too many people adopt that sort of reasoning. In a way, it is like getting on a bus without paying for your ticket. If most passengers pay, the bus can circulate and operate even for passengers without a ticket. But when the percentage of nonpaying passengers exceeds a certain limit, the public transport company is forced to cut routes and timetables, with a poorer service for all, whether paying or not.

In countries like Italy, vaccination rates, which had partially recovered compared to the 1990s, again began to drop from the first decade of the new century. In particular, adherence to the MMR vaccine fell from around 94 percent in 2010 to just over 85 percent in 2015. This was one of the lowest rates in Europe, well below the 95 percent required for herd immunity. At the same time, measles outbreaks rose to second place among European countries: 4,991 cases were registered in Italy in 2017, almost 6 times those of 2016, with 4 deaths and serious complications recorded in 35.8 percent of cases. This was reported by a Ministry of Health bulletin, which also stated that in 2017, the country had the second highest recorded cases of measles in Europe, after Romania.

This worrying situation did not escape the notice of the WHO, which drew attention to how Italy was becoming a problem for the entire European Union. As a result, in the summer of 2017, the Italian government was forced to make a certain number of vaccines compulsory as a condition for a child to attend school. The law has proven so effective in driving up vaccination rates that France and Germany, with similar problems, passed analogous laws in 2019.

A society where most people are vaccinated also helps protect those who cannot get immunized for medical reasons. This was the case of Rhett Kravitt, a seven-year-old boy in California who, in February 2015, had just recovered from leukemia. Rhett had been unable to return to school because of the risk that some unvaccinated children there could pass on diseases against which he lacked immunity. Rather than give up, Rhett and his family decided to talk to members of the school's board of directors, the Reed Union School District, north of San Francisco. He asked that the other children in the school be vaccinated, otherwise, with his weakened immune system from leukemia, he would be forced to study alone at home. One result of this campaign, which received wide media coverage, was that California changed state laws on vaccinations for those who wish to attend public-funded schools.

A world where people are more aware of the protection and safety offered by vaccines, a world with no need for mandatory requirements would, of course, be preferable. For now, however, laws reintroducing mandatory vaccines have been a necessary measure to guarantee public health in those countries.

## The Burioni phenomenon

"The Earth is round, gasoline is flammable, and vaccines are safe and effective. The rest are dangerous lies." It was with these sharp and clear-cut words that Roberto Burioni began his media career in the defense of vaccinations, in 2016 (Figure 3). Roberto Burioni, a professor of virology at Vita-Salute University in Milan, Italy, pronounced these words during a television program where the other panel guests were two local celebrities, who for years had spread false information regarding the alleged danger of vaccines.

In 2015, Burioni had started writing posts on Facebook, debunking myths about vaccines. He was particularly worried that children of his daughter's age,

Figure 3: Professor Roberto Burioni, an Italian university professor of virology who publicly defends vaccines and promotes evidence-based medicine on social media and TV (Wikipedia).

at the time in primary school, might not be protected from infectious diseases due to the amount of fake news being spread online.

To Burioni, it appeared somewhat incongruous for the state-owned television to give ample airtime to unfounded and dangerous information about vaccines, while that very same state was trying to contain the drop in vaccinations through its Ministry of Health. All of this was, for Burioni, a strong enough reason to start unmasking the false prophets of "no vax" movements on the Internet and on social media.

He started doing it, and continues to do so, with simplicity and directness, choosing very clear words and examples understandable to all. For instance, he has often repeated: "I know something about vaccines, viruses, and bacteria because I have studied them all my life, but I have no idea how to make a cake or how to fix a lamp, things for which I go to the bakery or call an electrician." For Burioni, the problem begins when anyone without a specific training and expertise feels entitled to talk about vaccinations and to ignore the opinion, based on evidence, of those who in fact do have the proper training and expertise.

The Internet and social media of course amplify the problem: without an editorial filter, the lesser expert voices are often lost, if not downright reviled, when compared to a multitude of nonexperts, to the detriment of competence. "Does an aeronautical engineer ask passengers to vote on how many wheels an airplane should have?" Burioni asked. "No, the engineer is the expert: he is trained for the job and it's his job to make that decision."

Things are not much better with traditional media such as television, where a misconception of equality necessitates the presence of two conflicting opinions, regardless of whether they are supported by evidence, facts, and demonstrable proof. For Burioni, it is impossible "to support a world where lies are conferred the same dignity as truth."

Not everyone appreciates Burioni's frankness. Some of his speeches have been criticized for being too harsh, further enraging an already highly socially divisive conflict on vaccines. More than once he and his family have received death threats.

However, several public health experts, not only Italians, credit him with changing the tone of the ongoing debate on vaccinations in Italy (*Science* magazine dedicated him a profile in the January 2020 issue). Burioni's interventions are highly popular, even when he is at his most blunt: on Facebook, Twitter, and on his "Medical Facts" website, he has more than one million followers. Without his efforts, it is unclear whether Italy's health authorities would have had the courage to make urgent decisions such as reintroducing compulsory vaccines for diseases, like measles, that had caused the numerous outbreaks mentioned earlier, due to families choosing not to immunize their children.

## Getting vaccinated is better

Ultimately, not vaccinating children is far riskier than having them vaccinated. It amounts to a lifelong exposure of their non-immune organisms to possible infections, some life-threatening. The choice is all the more relevant in today's world, where intense and widespread travelling brings people from all over the world into continuous and close contact. If people come from places where infectious diseases are still endemic, and they are not vaccinated, these diseases risk being reintroduced in places where they are considered problems of a distant past, their disappearance taken for granted.

Also, many decisions against vaccinations are not taken personally but are made on behalf of other individuals, children, who are too young to choose for themselves or have a say in the matter.

A perfect example of this last point is the story of Amy Parker (Figure 4). As a child raised in England in the 1970s, she fell ill with measles, mumps, rubella, meningitis, scarlet fever, whooping cough, and chickenpox: almost all diseases that are preventable with vaccinations. Breastfed for over a year, raised in the open air of the English countryside, she had a lot of exercise, and ate home-grown vegetables and no preservatives. Her mother believed that natural

Figure 4: Amy Parker as a child had many vaccine-preventable diseases. When she became a mother herself, she had her children immunized against every disease for which there is a vaccine. She has been telling her story publicly for some time, to convince other parents not to make the mistake that her mother had made with her (Amy Parker).

immunity and a lifestyle in harmony with nature would be enough to protect her child against infectious diseases, without the need for vaccines. Unfortunately, the facts have shown that this is not the case and when Amy Parker in turn became a mother, she had her children immunized against all possible diseases preventable with vaccinations. She also recounted her experience to "Voices for Vaccines," a parent-led association that supports and advocates for immunizing children on time, in keeping with the vaccination schedule set by public health authorities to reduce the incidence of vaccine-preventable diseases.

Sometimes kids do not even know that they have not been vaccinated. Some public health officers in Italy contact unvaccinated kids after their eighteenth birthday for follow-up, a practice adopted to provide these young people with an opportunity to be vaccinated. In the course of these follow-ups, some of these kids learn for the first time that they had never been vaccinated, nor have they been informed by their parents of this major health decision.

Vaccinating children is in fact an excellent idea: it is like taking out a health insurance policy against infectious diseases on their behalf, at excellent market conditions.

Let us hope that future parents, and adults in general, will rediscover the benefits offered by vaccinations, without necessitating an awareness imposed by new and widespread epidemics of diseases that are today preventable by immunizations.

By the way, vaccines have blazed their way into the twenty-first century brilliantly: safer, smarter, and more effective than those available in the past. However, their progress, which we will discuss in the next chapter, has come about much faster than an evolution in our way of thinking, often struggling in the face of novelty. It is time for our mental "hard drive" to reboot: we have to abandon all unfounded fears and instead seize the many excellent opportunities made available to our health.

The expression "adult and vaccinated" was once common, at least in Italy. It was a brisk and apt way of instilling confidence in the future: immunizations were offered to children who, after finishing high school and coming of age at eighteen, would be equipped to face the world in safety and with freedom. Nothing quite like vaccination guarantees the freedom to move, study, act, and work to build one's future, without the fear that a brother, a friend, a neighbor, or a colleague may transmit some terrible disease. And without the anxiety that social stability and economic prosperity may suddenly fall apart from an invisible and highly threatening creature.

In the last two years, with the COVID-19 pandemic still ongoing, we just got a tangible taste of this anxiety of contagion, of fear of economic recession, of limitations to freedom, and ultimately of what it means to live without vaccines.

In 1988, the beloved children's author Roald Dahl discovered that many parents did not wish for their children to be immunized. He wrote them an open letter in which he explained how he had lost his seven-year-old daughter Olivia to a fatal complication of measles, in 1962, when no vaccine was available against this disease. The letter ended with a resounding "LISTEN TO THIS." Yes, it is important that we all listen to this, before throwing away all the well-being, health, and freedom that vaccines so generously and silently give us.

# 4
# Chasing Moving Targets

From the 1980s onwards, vaccines began experiencing a difficult time. In addition to the growing confusion that was spreading among the public, as we have seen, many scientists were redirecting their research towards different objectives, although for quite separate reasons.

Most of the vaccines in use in the last century were developed thanks to the application of Pasteur's principle "isolate the germ, kill it, inject it." For a century, vaccines were produced along this principle, and are still effective today against most pathogens that continue to manifest themselves "in the same clothes," so to speak.

However, several diseases caused by microorganisms can often disguise themselves or remain hidden. These are also serious and widespread diseases, such as influenza, pneumonia, meningitis, AIDS, malaria, TB, and hepatitis C. They are caused, respectively, by the influenza virus, *Pneumococcus* and meningococcal bacteria, HIV, *Plasmodium* parasites, *Mycobacterium TB*, and HCV.

In the 1980s, obtaining vaccines against these "moving targets" was considered a "mission impossible." In the meantime, traditional methods for producing vaccines against bacteria and viruses that had remained stable over time had run dry.

While conventional technology for developing vaccines was in danger of becoming a blunt weapon, the same could not be said for infectious diseases, who had lost none of their vigor. On the contrary, the death rate from infections, which had seen a decline during the second half of the twentieth century, had begun to reverse its trend towards the end of the century and was continuing to rise (in the United States, the increase was roughly 58 percent from 1980 to 1992).

Over that same period, infectious diseases continued to represent the leading cause of death worldwide, with AIDS and hepatitis C appearing as the new emergencies. Infections such as TB, considered under control at least in the richest parts of the world, were now reemerging in hospitals in developed countries.

Other threatening outbreaks—avian flu and influenza A, SARS, MERS and COVID-19, West Nile, Ebola, to name but a few—have also cropped up in the new century. Last but not least, even trivial diseases, such as sinusitis and bronchitis, have become increasingly difficult and expensive to treat, because the agents that cause them have developed an increasing resistance to drugs.

In short, the need for new vaccines has never been greater. In the 1980s, however, many scientists abandoned the field to devote themselves to areas of research that looked more promising. They were discouraged by technical difficulties that, at first sight, seemed insurmountable, and perhaps also by the low popularity of vaccines among the public. Despite these difficulties, progress was happening.

## Recombinant vaccines: The case of hepatitis B

Hepatitis B is a viral liver disease, which spreads through infected needles, transfusions with contaminated blood, and through sexual contact. The infection can also be transmitted from mother to child during childbirth or between children through cuts, bites, or scratches.

According to a WHO estimate, some 2 billion people—roughly a quarter of the world's population—have become infected with the virus at some point in their lives. Of these, a few hundred million are chronic carriers of the disease, often without even knowing they are contagious. In 2015, again according to WHO, 257 million people were living with chronic hepatitis B, and in that same year the disease caused roughly 887,000 deaths, mainly from cirrhosis and liver cancers brought on by the infection. Cases of chronic hepatitis develop mostly among children who have encountered the virus under the age of 5, and about 1 in 5 of them will die prematurely from cirrhosis or liver cancer.

In 1976, at the Merck Institute for Vaccinology in West Point, Pennsylvania, Maurice Hilleman and his colleagues purified a viral antigen from the plasma of infected people: it was a protein present on the surface of the hepatitis B virus (HBV).

Short terminology break: the word *antigen* indicates any foreign substance that the immune system is able to recognize in a specific way, while *plasma* is the liquid part of the blood.

It turned out that the animals injected with the purified protein became immune to the disease: the antigen worked as a vaccine. However, it could not be mass-produced, because the number of infected people was insufficient for extracting enough antigen from plasma.

The solution arrived in 1982, when William Rutter and Pablo Valenzuela, at the University of California at Berkeley, induced yeast cells to produce the protein. This feat, extraordinary for those years, was accomplished by inserting the protein's gene, obtained from viral DNA, inside the yeast's chromosome.

This was one of the first applications of recombinant DNA technology to vaccine development. The very convoluted techniques of the time were used to make a kind of "cut and paste" technology with pieces of DNA from different organisms. The viral protein, produced in large quantities in those fertile bio-factories that are yeast cells, appeared to have the same properties as the antigen found in the plasma of humans infected with the virus.

From those yeast cells, the first recombinant vaccine in history was born. This vaccine was no longer made from an entire, inactivated viral particle. It was instead derived from a single viral gene which, despite being extracted from its original habitat and inserted among foreign genes, was still able to produce an effective antigen against the hepatitis B virus.

Since 1992, the WHO has recommended the use of this vaccine in all countries where the disease is widespread. By 2019, HBV vaccinations in the world had risen to 85 percent, compared to around 30 percent in 2000. Thanks to the vaccine, the percentage of children in many countries who contract chronic hepatitis has decreased from 8–15 percent to less than 1 percent. However, global statistics conceal large differences: in Africa, for instance, the coverage of infants with this vaccine is still just 6 percent.

Increased use of the vaccine would result in more countries benefiting from a reduction in liver tumors caused by infection with the HBV, something which we have seen decrease significantly among vaccinated populations. In fact, we could say that the shots are killing two birds (HBV infection and any resulting liver cancer) with one stone (the vaccine). For this reason, the anti-HBV vaccine can be considered the first anti-tumor vaccine, as well as the first vaccine derived from recombinant DNA technology.

## Computer-designed vaccines:
## The case of whooping cough

Pertussis, or whooping cough, is a respiratory disease caused by bacteria that live in the mouth, nose, and throat. Over 24 million cases of pertussis were reported in 2014, with more than 160,000 deaths in children under 5, the majority in Africa, according to a WHO estimate. A vaccine against pertussis had been developed and approved for use in the 1940s, from the killed cells of *Bordetella pertussis*, the bacillus responsible for the disease.

Vaccinations against pertussis in the second half of the twentieth century had contained the spread of the disease quite successfully, although as a result of some serious side effects, the vaccine was limited to a minimum dosage, sufficient only to protect children in the first months of life, when the disease is more dangerous. Taking advantage of this limited use of the vaccine, the bacillus of *Bordetella pertussis* continued to circulate undisturbed. A safer vaccine was needed.

The ideal candidate for the vaccine was pertussis toxin: a poisonous protein that is produced by the bacterium. The toxin can be recognized by the immune system, which then generates a protective response against the disease. However, the toxin had to be rendered harmless before it could be used. The traditional Ramon method, based on formaldehyde (mentioned in Chapter 2), did not work well in this case: once the toxin was chemically inactivated, it caused an insufficient immune reaction, requiring large doses for effective immunization.

In the late 1980s, researchers at Sclavo Institute, in Siena, Italy, tried a new approach. After all, the researchers reasoned, the toxin is a protein, and like all proteins it is made up of a series of building blocks—amino acids—their order determined by the sequence of their corresponding gene. If, by changing some amino acids, it was possible to obtain a nonpoisonous toxin that was otherwise identical to its natural counterpart, this might induce a reaction within the immune system.

Researchers at Harvard University had already done this for the diphtheria toxin, introducing random mutations in the toxin gene. The Harvard test was a good start: it showed that, in principle, a nonpoisonous but immunologically active toxin could be obtained by modifying its gene. The next step forward was to replace the randomness inherent in this method with something more predictable and reproducible.

In Siena, the researchers began studying the toxin's structure with the aid of a computer. Thanks to some specially designed software, they were able to build a three-dimensional model of the protein and observe its structure as against other known proteins. In this way they were able to discover that only two amino acids, out of a total of over eight hundred, were essential for conferring toxicity to the molecule. If those two amino acids could be replaced with others, the scientists reasoned, the toxin should become harmless.

Separate software helped identify the nucleotides for the two amino acids in the toxin gene. A modified gene and its nontoxic protein were then obtained by targeting mutations for the nucleotides. Although harmless, the protein was otherwise identical to its natural form, especially with respect to its immunological properties. A new vaccine against pertussis was born, far safer and more effective than its predecessor.

This was the first time that a vaccine had been designed by starting off from a theoretical hypothesis, elaborated with the aid of a computer, and then confirmed by laboratory experiments. What was the benefit? The new vaccine against pertussis contained only three basic elements: the mutated toxin, studied and modified with a repeatable and non-random procedure, and two additional antigens purified from the bacterium. The hundreds of molecules normally contained within an entire deadened bacterium were no longer present in the vaccine.

## Conjugate vaccines: The case of *hæmophilus*

*Hæmophilus influenzæ* is a bacterium that can live in our body without causing any problems, until a viral infection or a lowered immune function creates an opportunity for its proliferation. Microorganisms that behave in this way are called opportunistic pathogens. When the *hæmophilus* finds an opportunity to reproduce, it can cause meningitis, especially in children. Meningitis is a serious and sometimes fatal infection of the membranes that line the spine and brain.

Precisely this manner of infection, which takes advantage of the opportunities provided by other diseases, is the reason for the misleading name of this bacterium. *Hæmophilus* was, in fact, identified for the first time in 1892 during a flu epidemic, and for the next forty years was considered to be the causative agent behind the flu, until the first influenza virus was discovered in 1933.

There are six recognized strains of *hæmophilus* (a, b, c, d, e, f), but only type b (or *Hi*b) is considered capable of causing meningitis and other childhood diseases, which is why vaccinologists have focused their attention on it.

An obstacle that has long hindered any attempt to develop a vaccine is the fact that the bacterium is coated in a capsule of complex sugars, or *polysaccharides*. Other bacteria, such as *pneumococcus* and *meningococcus*, also use this trick to elude the immune system. In this way, the polysaccharides for *hæmophilus* can escape the T lymphocytes' surveillance system, but they cannot bypass that of antibodies and complement (to learn more about both T lymphocytes and complement, you can read the Appendix on our defenses and how they work). But while this partial control occurs only in adults, providing a short-term immune response (you can contract *hæmophilus* more than once), it does not protect children, an age group with the highest incidence and mortality rates for *Hi*b.

Since the 1930s, it has been known that polysaccharides provoke a greater immunological response and immunological memory when the sugar is chemically bound to a protein. It is generally rare for a truly effective vaccine to contain only one isolated constituent of the pathogen, because a more effective protection is obtained if several components of the immune system are stimulated simultaneously.

T lymphocytes recognize antigens only if they are made up of proteins. Based on this observation, four American researchers—Porter Warren Anderson, of the University of Rochester; David Hamilton Smith, of the David H. Smith Foundation; John Robbins and Rachel Schneerson, of the National Institutes of Health (NIH)— developed the first conjugate vaccine against *Hi*b, consisting of a polysaccharide bound to a protein.

The *Hi*b conjugate vaccine has been available since the late 1980s, and by 2016 it had already been introduced in 190 countries. In the United States and other Western countries, the incidence of *Hi*b meningitis has been reduced by more than 99 percent, a result never achieved in such a short time by any other vaccination campaign.

The four scientists who developed the first conjugate vaccine against *Hi*b were awarded the Lasker Prize, a recognition that on more than one occasion has been the antechamber to the Nobel Prize.

Another conjugate vaccine, developed following this same principle against Men C, has effectively eliminated the disease caused by this bacterium in the United Kingdom.

## The genomic revolution

The Human Genome Project was initiated in 1990, as an extensive scientific undertaking involving hundreds of laboratories across eighteen countries. The Project's goal, coordinated by the US Department of Energy and the US NIH, was to analyze the genetic material—or genome—that determines the fundamental characteristics of each individual. This meant identifying all human genes, of which we have about 23,000, and make their sequence accessible to all researchers in the world, so that the function of each gene could be studied further. The idea was that from this massive work, it would be easier to understand how our body works, while medicine would acquire new tools for more precise diagnoses and more effective treatments.

The Project's scope was impressive: more than three billion nucleotides— the "building blocks" of DNA that make up the genes—had to be read one by one, on each of the twenty-four chromosomes found in the cells of our species (twenty-two chromosomes, common to males and females, plus the two sex chromosomes X and Y). The estimated time for the Project was fifteen years.

At first, hardly anyone believed that a challenge of this magnitude could be completed in such a short time. Incredibly, automatic technologies were invented with unprecedented speed to build gene maps and analyze the DNA sequences. The time to decipher a single gene, which in the 1980's took a year at least, was now shortened to days or even hours.

The acceleration was extraordinary, and for biologists it was like jumping directly from horse-drawn carriages to airplanes. Private laboratories such as Celera Genomics, directed by the American biologist Craig Venter, also contributed to the race against time.

The amount of data on genes and DNA sequences that was coming out of the labs involved in the Human Genome Project began to increase exponentially. This created a vital need for tools to analyze, order, and catalogue those sequences. Information technology rose to the occasion with powerful computers and programs that, in a few minutes, could reassemble a huge mass of genetic fragments in their proper order.

Altogether, the new technologies accelerated the completion of the Human Genome Project, which ended two years earlier than expected, in 2003. It was exactly fifty years after James Watson and Francis Crick had proposed the double helix as a three-dimensional structure for the DNA molecule.

Those vaccinologists that had not deserted the field became very excited by the genomic revolution. Many of them sensed that with the innovations put forward by the Human Genome Project, a new era would also open for the development of vaccines.

The first fully deciphered bacterial genome, belonging to *hæmophilus*, dates back to 1995. As of February 2022, the US National Library of Medicine database alone contains information on more than 21,000 genomes for eukaryotic organisms, 390,000 genomes for prokaryotes, and 47,000 viral genomes, with numerous variants for each species. For the HIV virus alone, which causes AIDS, the Los Alamos National Laboratory database contains more than 950,000 sequences, of which 16,000 are complete. As of February 2022, 8.7 million sequences of SARS-CoV-2 virus have been deposited in the GISAID database, of which 8.5 million are complete. And the numbers are increasing every day. These are impressive figures, considering that every single genome can include thousands of genes and that, at the turn of the century, microbiologists spent most of their time "reading" one gene at a time.

Why is it useful to know the full genome of so many bacteria, viruses, and parasites? Prior to genomics techniques, it was impossible to diagnose some diseases using the conventional tools of microbiology. The HCV, for instance, could not be grown in the lab. Until 2016, the virus could not even be observed under an electron microscope, despite the fact that its complete genomic sequence had been available for more than twenty-seven years. Fortunately, since 1989, knowing the virus' genomic data has made it possible to verify its presence in blood and in blood derivatives, in this way eliminating the risk of viral transmission through transfusions.

Another example is the SARS epidemic, which in just five months, from February to July 2003, caused over eight thousand infections and killed nearly eight hundred people. The genome for the coronavirus responsible for SARS was decrypted and deposited in public databases within a month of it being isolated. The time needed to obtain the complete genome sequences of human pathogenic microorganisms that have appeared since then—in particular the 2009 influenza A virus and SARS-CoV-2, in 2019—were now reduced to a matter of days. These extremely fast procedures demonstrate how new technologies allow for almost real time action even with novel and never-before-seen infections.

The entire deciphered genome has also proven extremely useful for developing new vaccines, as in the case of the meningococcal B (Men B) vaccine. This challenge, considered impossible for more than a century, can be regarded as one of the most important revolutions in the world of vaccinology since the time of Pasteur.

## A revolutionary strategy against meningococcal meningitis

Beatrice, or Bebe, Vio, born in Venice, Italy, in 1997, was struck by meningitis when she was eleven-years-old. The disease spared her life but left her a punishing legacy (Figure 1). Her biological legs and arms were partly destroyed by bacterial type C meningitis, an unforgiving disease that made it necessary to amputate under the knees and elbows of all four limbs. After a hospitalization that lasted 104 days, and long periods of psychomotor rehabilitation and physiotherapy, Bebe resumed her fencing training, a sport she had been practicing before her disease, from a wheelchair. Thanks to a special prosthesis created specifically to allow her to hold the foil, since 2014 she has gone on to become Paralympic World Champion several times over.

Worldwide, one person dies of meningococcal meningitis every eight minutes. For every ten cases, one is fatal and the disease, which is not always recognized

Figure 1: The Italian Paralympic athlete Bebe Vio (photo by Augusto Bizzi).

and diagnosed correctly, can kill within twenty-four hours from the onset of infection. One in five people who recover risk bearing the consequences of the disease for life, including kidney failure, brain damage, loss of limbs, and loss of hearing. Children and adolescents are the most vulnerable.

Meningococcus (its scientific name is *Neisseria meningitidis*) is a regular host of the bacterial flora in an adult's nose and pharynx. It typically goes unnoticed, except when it finds an opportunity to reproduce, usually once the immune defenses are either weakened or busy fighting another infection. This makes for a pathogen that results in a widespread reservoir of asymptomatic people, from which (unfortunate) cases of disease emerge from time to time.

We know of six different types of meningococcus (A, B, C, X, Y, and W135), classified according to the chemical composition of their complex sugar capsule.

Only type A causes large-scale epidemics, mainly in Africa and Asia, while in Europe and America most cases are due to types B and C.

Vaccination with the type C conjugate vaccine has resulted in the elimination of the disease in the United Kingdom (see Chapter 3), while in the early 2000s, Men B was still in circulation. Today, the diseases caused by five of the six serogroups are preventable by vaccines, with the most recent vaccine now developed against Men B.

Men B has a sugar capsule, initially giving rise to hopes that a conjugate vaccine might work against it. The idea turned out to be impractical because the Men B capsule leaves the immune system, which works well against other types of capsules, completely indifferent. The reason for this immune inertia lies in the fact that the Men B is a particularly cunning bacterium: it does not just hide under any sugar disguise, it chooses a sugar molecule from the human repertoire that is present in many of our cells. It can in this way circulate freely within our organism, escaping our immune surveillance system, which is led to "believe" that it is in the presence of something that is ours.

This all made it impossible to build a conjugate vaccine against Men B based on the capsule sugar. The vaccine would have stimulated the immune system to react against the sugar, mistaking it for a harmful substance, which would have created considerable problems for the bacterium, but with disastrous consequences for human cells, tissues, and organs containing that same sugar. In other words, there was a risk of an autoimmune reaction (more information on autoimmune diseases can be found in the Appendix on our defenses and how they work).

Other candidates for a Men B vaccine, using a procedure that had become standard since the 1980s in conventional vaccinology, were some antigens present in the bacterium's outer membrane. In fact, a vaccine built with one of these proteins made it possible, in 2004, to eradicate the disease in New Zealand, a country where meningitis epidemics had become increasingly frequent. But the New Zealand's outbreaks were (thankfully) due to a single strain of Men B, against which the purpose-built vaccine was very effective.

But the vaccine used in New Zealand would not be as effective in other parts of the world, where hundreds of different strains of Men B are often present at any one time. The differences between the strains are caused by the presence of more than seventy proteins found in the outer membrane.

Men B's capacity for variation is remarkable, making Johann Sebastian Bach seem an amateur by comparison. Most vaccinologists considered the mission to be impossible. One exception was the group in Siena, which had meanwhile become a branch of the American biotechnological company Chiron. They decided to challenge Men B with a radically new approach.

Men B, they reasoned, is too cunning and versatile to consider the idea of developing a vaccine based on the few known antigens only. So they decided to examine all the proteins that Men B could produce to see if they could be recognized by the immune system to induce a bactericidal response. The inventory of all Men B proteins is located in the genome, more specifically, in the *pangenome*, i.e., the set of all genomes for all known strains of Men B.

To analyze all the genomes, the Siena group co-opted Craig Venter, the scientist involved in the Human Genome Project, and his contribution would be crucial. Thus began the Sienese bet against one of the most elusive bacteria in creation.

## The impossible challenge overcome by vaccinologists

To master the anti-Men B vaccine, scientists had to pass a series of tests, a bit like Tamino must do in Mozart's *The Magic Flute* in order to free the beautiful Pamina from the evil Queen of the Night. Except that, instead of Tamino's magic flute, the Sienese scientists possessed some powerful computers.

Let us take a look at the steps they followed.

### 2,000 theoretical proteins

Each Men B strain's genome is about 2.2 million nucleotides long and can theoretically produce over 2,000 proteins. How could they all be analyzed within a reasonably short time? Thanks to computers that sped up the analysis, the time required was significantly reduced.

### 600 membrane or secreted proteins

Another software helped identify 600 gene sequences, corresponding to just as many proteins, still unidentified and probably located on the surface of the bacterium, or released by Men B into the host organism. But how many surface proteins can, in fact, produce those 600 gene sequences? The theoretical predictions made by the computer had to be verified with laboratory experiments.

### 344 purified proteins

In the lab, each gene sequence was individually inserted into the chromosome of a single *Escherichia coli* bacterium. Perhaps the most common inhabitant of the human intestine, *E. coli* is among the most used bacteria in biology because it grows well in the lab and it is easy to introduce foreign DNA into it. The idea behind this experiment was that, when one of the meningococcal sequences were inserted into it, the bacterium would produce the corresponding recombinant protein wherever possible. Using this method, 350 proteins were obtained from the 600 gene sequences introduced into *E. coli*, of which, 344 could be purified.

### 91 immunogenic proteins

The next step was to understand which of the purified proteins would be recognized by the immune system. Each protein was injected into a mouse, and its serum was then analyzed for antibodies developed against that specific protein. This led to antibodies being found in the mice sera, corresponding to 91 proteins, recognized as surface antigens by the immune system.

### 28 bactericidal antigens

Of the 91 antigens located on the external surface of the bacterium, 28 were able to induce a bactericidal activity by the mouse's immune system.

**5 antigens conserved by evolution in all strains**

A further check ascertained that 5 of the 28 antigens had the optimal qualities for a vaccine: they were located on the surface of the bacterium, and they were present in most of the strains of Men B, likely conserved in evolution, and were capable of inducing a universal bactericidal response against all known strains.

## An interesting discovery

One of the common antigens to all Men B strains was a protein called GNA1870, which naturally occurs on the surface of the bacterium. It was discovered that it was this very protein that kept Men B from getting noticed by the immune system. This is because GNA1870 binds to factor H, an essential component of one of the immune system's most important surveillance mechanisms. Thanks to the link with the factor H, Men B passes itself off as belonging to the human body and can survive undisturbed in the bloodstream. Now, the interesting thing is that the GNA1870 protein only binds to the human factor H, but not to the mouse or rat equivalent. This explains why scientists have never been able to study the infection caused by Men B in a laboratory animal.

Once the clinical trial in adults and children was completed, the Men B vaccine was approved by regulatory authorities in 2012, after about twenty years of studies and research. Today, it is used in many countries.

## Reverse vaccinology, or the art of making vaccines backwards

In the last fifty years, the main ingredients in vaccines were specific antigens (such as proteins or sugars) that were stably found on the surface of bacteria and viruses. As we have seen in this chapter regarding the hepatitis B vaccine, researchers first had to isolate an antigen, then discover its functions and retrieve its corresponding gene sequence. All this had to be done before the gene could be introduced into a biological factory (brewer's yeast, for instance) that could manufacture the antigen in industrial quantities.

Besides requiring a very long time, this process was also quite risky because all the "bets" were on a single "horse." If, in practice, the chosen antigen failed to work well as a vaccine, the researchers would have to literally start again from scratch.

Reverse vaccinology bypasses this lengthy procedure, as well as reducing the risks attached, by taking the opposite path: researchers no longer proceed from one antigen to a gene, they begin instead by reading the entire genome in order to identify all possible antigens (Figure 2). The number of antigens identified in this way is far greater than what was possible under the traditional approach.

To build a vaccine today, scientists no longer bet on the few antigens they encounter more or less by chance. Instead, they can now examine every single candidate that can be predicted by analyzing the genome.

But even with the aid of a computer's processing power and analysis, antigen selection is still very difficult: in the case of Men B, as we have seen, only five antigens were ultimately considered suitable for the vaccine, from over two thousand theoretical gene sequences.

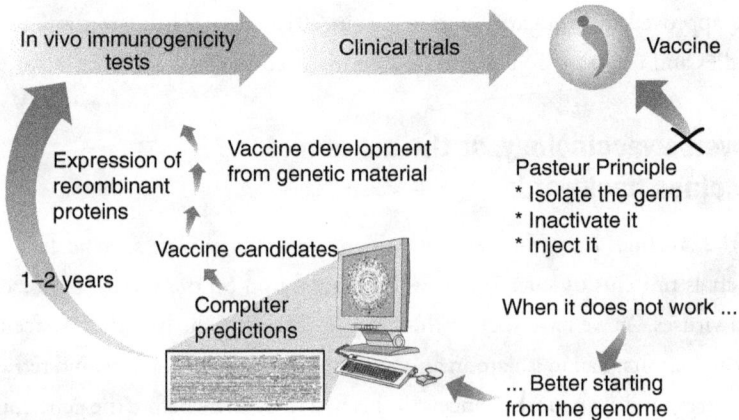

Figure 2: With reverse vaccinology, vaccines are designed starting from the analysis of the genome with the aid of a computer software.

Still, thanks to computer analysis and lab experiments, it was possible to reject as many as 1,995 candidates: molecules that the immune system would not have recognized, or antigens that could present safety issues.

In other words, reverse vaccinology allows for the development of vaccines that are far safer and more accurate than the earlier quite safe preparations produced in the previous century using Pasteur's principle.

Using a rather audacious comparison, we could say that the revolution brought about by reverse vaccinology is similar to the one that took place in the world of sports betting. Prior to computers, people would place a bet on their favorite team for emotional reasons, or because they believed a team would win. Then came computer science, and with it all the statistics on the various teams' past performances, which introduced slightly less random elements of probability. From that time forward, we can bet not only on our favorite team, but also on those games that more rational estimates foresee as winners. Vaccinology is (thankfully) a science and is (thankfully) much more precise than football statistics, but the approach used by researchers has changed in a similar way.

## Reverse vaccinology 2.0 and vaccines for pregnancy

An even more recent wave of new technologies in the fields of structural biology, human immunology, and genome editing is providing invaluable information and tools for the design of vaccines that were considered impossible to develop.

Let us take, for instance, the respiratory syncytial virus (RSV), a fairly common pathogen that usually causes mild, cold-like symptoms. Most people recover in one or two weeks, but for infants and the elderly, the infection can cause a more serious, at times fatal, illness. Around the world, this virus is second only to malaria for the number of deaths among children aged one and under.

Since the 1960s, several attempts have been made to obtain a vaccine against RSV using the classic methods of vaccinology. Unfortunately, these attempts were unsuccessful because some crucial information was lacking about the structure assumed by the viral proteins that are most recognized by the immune system during the infection.

The virus penetrates into the host cells through the linking of a particular protein, glycoprotein F, to the cell membrane. Glycoprotein F can assume one of two forms: one before and one after the fusion of the viral envelope with the cell membrane. Jason McLellan, Barney Graham, Peter Kwong, and their colleagues at the NIH in Bethesda, Maryland, studied both forms of glycoprotein F, atom by atom, using the methods of structural biology. Thanks to X-ray crystallography, they were able to obtain a crystal for each protein, bound to a neutralizing antibody (Figure 3).

Short terminology break: a *neutralizing antibody* is a protein that, by binding to another protein belonging to a bacterium or a virus, interferes with the

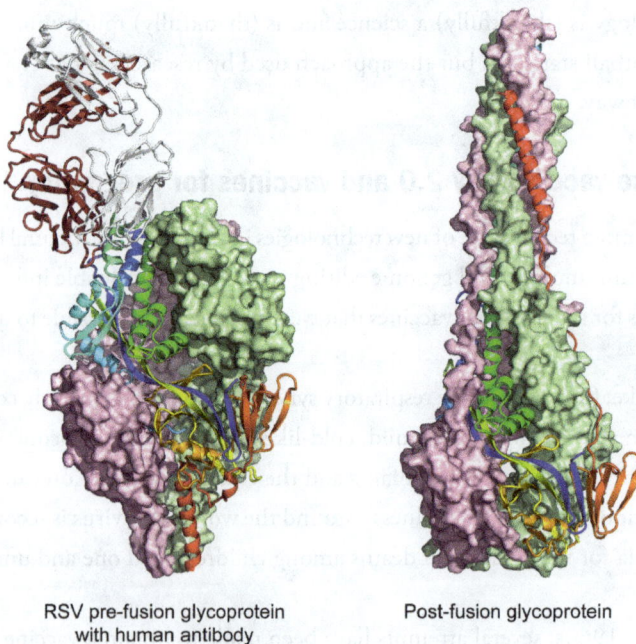

RSV pre-fusion glycoprotein
with human antibody

Post-fusion glycoprotein

Figure 3: The F-glycoprotein of the RSV virus is shown, on the left, in the state before fusion with the cell membrane, while bound to an antibody, and, on the right, in the form it takes after fusion (NIAID).

latter's ability to infect a cell, rendering it harmless. Many antibodies are produced by the immune system after a microorganism has infected the host, but few are neutralizing.

From the crystallography, the researchers discovered that the most effective vaccine would be one focusing on the form assumed by glycoprotein F before it enters the cell, because it is the one best recognized by the immune system. But they still needed to overcome another obstacle: this form of glycoprotein F is highly unstable.

Thanks to structural analysis, the researchers were able to not only understand the nature of the problem, but also identify which of the protein's amino acids could be modified in the lab—through gene mutations—in order to make their structure more stable. Using this method, in 2013 it was finally possible to obtain a vaccine capable of inducing a particularly effective immune response, at least in the preliminary stages of the clinical trials. If these results are confirmed in larger studies, an effective and accurate vaccine against this disease will finally become available.

Structural vaccinology is the design of vaccines, such as the one against RSV, and is based on the study and modification of single atoms within the structure of viral proteins. This approach is becoming increasingly common and useful to circumvent problems previously hindering the prevention of several infectious diseases. As we will see in Chapter 8, these same researchers and their methods have been crucial in selecting the antigens for COVID-19 vaccines.

The approach is part of what vaccinologists call reverse vaccinology 2.0, the second phase of the far-reaching revolution that has radically changed the way new vaccines are created and developed. Other discoveries that have contributed to these advances have occurred in immunology, with the increasingly sophisticated ability to do lab research on B lymphocytes—those human cells that produce antibodies—and follow their evolution.

In nature, an organism reacts to an infection by initiating production of B lymphocyte clones, that is, many immune cells that produce the same antibody. Different people produce different clones, giving rise to different immune responses, more or less powerful, rapid, and effective. Today, with monoclonal antibody technology, we can produce large quantities of a specific antibody in the laboratory to target a specific substance. This allows us to study the characteristics of many antibodies produced against the same pathogen by many B lymphocytes from a vast number of different people. The monoclonal antibody technique was developed in the 1970s by Georges Köhler and César Milstein, who were awarded the Nobel Prize in Physiology or Medicine for their efforts. By combining this technique with structural and computational biology, it is possible to obtain information regarding the structure of each antibody in its more or less lethal "embrace" with its own antigen. This helps researchers understand which match between the two molecules works best for our defenses, and allows us to use that information to design a vaccine with the best possible antigen in an optimal configuration.

Being able to study the vast repertoire of antibodies with their own antigens in atomic detail makes it now possible to design vaccines against those antigens, or parts of antigens, that can stimulate the most precise and desired immune response in a majority of people. Sometimes, to optimize the response, the structure of the antigen has to be modified, as we have just seen in the case of the vaccine against RSV. Today, it is possible to obtain variations in genetic material, and consequently in proteins, both quickly and with great precision, thanks to genome editing techniques such as CRISPR/Cas9, developed by Gabrielle Charpentier and Jennifer Doudna, the two Nobel Laureates in Chemistry 2020.

Thanks to reverse vaccinology 2.0, a vaccine against RSV has been developed that can be safely and effectively administered to women in the third trimester of pregnancy. Other vaccines for expectant women against pertussis, influenza, and type B *Streptococcus*, have been created to avoid diseases that can kill infants in their first four to five months, or leave them with lifelong consequences.

Seven out of ten deaths from pertussis occur in children under two months of age, when they are still too young to be protected by a vaccination directly administered to them rather than to their pregnant mothers. As for the flu, it is best avoided during pregnancy because expectant women are more likely to contract it in a severe form, while vaccination is the best way to protect the baby from the flu and its complications for several months after birth. Premature births and asphyxia at birth are often associated with type B *Streptococcus*, and babies born to mothers with this infection can experience sepsis, pneumonia, meningitis, and death.

If mothers are immunized against these pathogens during pregnancy, the antibodies that pass through the placenta protect newborns during the most critical stage, when these diseases can be fatal.

We have now come to the end of this chapter, where we have seen how computers and the most advanced know-how and technologies for molecular and structural biology, and immunology, have led to startling results, considered impossible just twenty years ago.

Computers and biotechnologies—it must be remembered—are human artifacts that are neither infallible nor magical; their calculations, methods, and procedures must be set up and used with care, and their conclusions carefully evaluated. That said, the contribution, the potential, and the opportunities that these tools provide to modern medicine and biology are both extraordinary and invaluable. Suffice it to say that the time it takes to discover and bring a new vaccine to clinical trials can now take less than a year, as we have seen during the COVID-19 pandemic. Before reverse vaccinology, at least fifteen years would have been needed.

# 5

# Vaccines Wanted Against Four Big Killers

As we have just seen in the last chapter, great progress has been made in the world of vaccines in recent years, a progress that was unthinkable just twenty years ago. Despite this, four serious infectious diseases are not yet preventable with vaccines. They are AIDS, malaria, TB and hepatitis C, which together kill more than three million people every year. It is as if more than a third of the inhabitants of London disappeared every twelve months.

About 690,000 people become infected every day with one of these diseases (that is more than the entire population of Bristol every 24 hours), adding to the nearly 2.2 billion individuals already infected: over one and a half times the population of China[1].

Despite the tremendous advances in vaccinology, we still lack effective vaccines against these great killers. Why is this?

Immunologists often argue that to overcome these pathogens we need a "paradigm shift." It is an elegant way of saying that we still know too little about these diseases—whether they are recent or only recently identified, such as AIDS and hepatitis C, as old as malaria, or a "comeback" disease like TB. We need to understand them more thoroughly if we want to find better preventive remedies.

---

[1] In 2020, it is estimated that 1.5 million people became infected with AIDS, with 690,000 deaths, bringing the total number of cases to more than 37 million. In 2020 there were 241 million estimated cases of malaria and 627,000 deaths. In 2020, about 10 million new cases of TB were registered, with an estimated 1.5 million deaths. Today, we can count around 1.9 billion people infected with TB worldwide. There are an estimated 1.5 million new cases of hepatitis C every year and there were nearly 290,000 deaths in 2019, while 58 million people worldwide are estimated to live with the infection (WHO and UNAIDS data).

The gaps in our knowledge mostly concern the immunity of the mucosal sites, through which most of the invaders pass; in the complement system, which should provide an initial internal barrier, but fails to do so; in antibodies, which struggle to recognize the antigens or do not know how to mobilize the next level of defense; and finally in the T lymphocytes, which fail to inflict the final blow on the pathogens.

The obstacles are not just scientific. These four diseases are concentrated above all in developing countries, where massive economic, political, and social problems combine with the difficulties of research. However, even if these unscientific complications were overcome, unfortunately, vaccines would still not be within reach. In this chapter we will try to clarify why this is.

## Why is there still no HIV vaccine?

The first cases of AIDS were reported in 1981 in the United States. Forty years later, we know a lot about AIDS. The disease is caused by HIV, a so-called retrovirus formed by an envelope of proteins and sugars and by RNA as genetic material. The virus was isolated at the Pasteur Institute, in Paris, France, by Luc Montagnier and Françoise Barré-Sinoussi, who received the Nobel Prize for their discovery in 2008.

HIV infects the T cells—a type of lymphocyte—in the immune system, destroying them and weakening the entire defense mechanism's function. However, it is rare for any signs or symptoms to appear at the onset of the infection. It is only later, as the infection progresses, that immunity deteriorates and those who are HIV-positive become increasingly vulnerable to other diseases.

AIDS generally manifests when the HIV infection has reached an advanced stage. Up to 10–15 years can elapse between the moment of contagion and the first signs of the disease; this interval can be even longer if the HIV-positive person is treated with antiretroviral drugs, available since 1992.

HIV can be transmitted through unprotected sex, transfusions of contaminated blood, and the sharing of needles and syringes. It can also pass from an HIV-positive mother to her baby during pregnancy, childbirth, and breastfeeding.

Despite the accumulated knowledge and the progress made in therapies, which have transformed AIDS into a chronic and no longer fatal disease, an effective vaccine against AIDS still does not exist, nor is there one on the horizon. Why is this?

HIV is a virus that is totally different from any other pathogen for which an effective vaccine has been obtained so far. As we have seen, diseases like small-pox, measles, and polio can also kill. However, many people recover from these diseases through a spontaneous immune response that eliminates viral particles and infected cells, through the action of antibodies and T lymphocytes. Also, with healing, a lasting immune memory is created.

With these diseases, nature had already "made the experiment" that vaccinol-ogists need to replicate to create a vaccine. The body's natural ability to defeat an infection is, indeed, a proof of principle that a vaccine can be obtained against a certain pathogen. After all, what is a vaccine, if not an artificial way of inducing the organism to do the same things that the immune system at times does alone?

But with HIV, this proof of principle does not exist. Since the beginning of the epidemic, of the approximately eighty million people who have become HIV-positive to date, not a single case has been documented of someone who has managed to eradicate the virus completely and spontaneously. And, indeed, several people have succumbed to a reinfection of the virus, which means that the initial infection does not protect against further bouts of the disease.

Since nature does not help us, "we must do better than nature." In an interview with *The New England Journal of Medicine*, Anthony Fauci described the chal-lenges of building the HIV vaccine. Director of the National Institute of Allergy and Infectious Diseases in Bethesda, Maryland, and one of the world's leading experts on HIV and AIDS, Fauci (Figure 1) has been an advisor to numerous

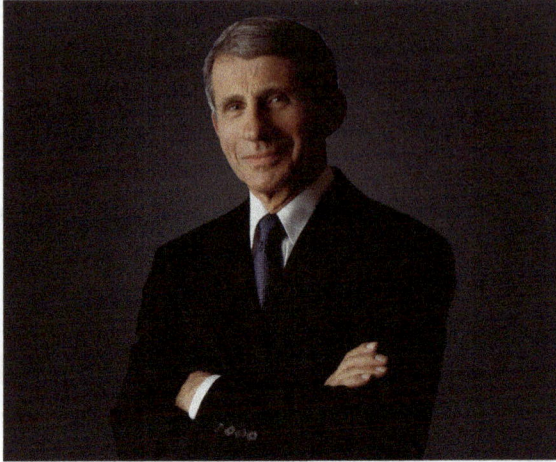

Figure 1: Anthony Fauci (NIAID).

American presidents during various epidemic crises, including COVID-19, when he was often interviewed on television and other media.

Why is the immune system unable to block HIV? When HIV infects a cell, the virus integrates its own genetic material into the host's DNA. Generally, the virus prefers *T helper lymphocytes*, the cells that coordinate the work done by many other of the immune system's agents and molecules. The virus then remains *latent* for a long time in some of these cells; in other words, it is almost inactive and invisible to the immune surveillance systems. This state of latency is established within days, or at most, a few weeks after infection, too short a time for the protective antibodies and other defense mechanisms to respond effectively.

With its own genetic material integrated into the genome of T helper lympho-cytes, HIV creates an infection that is, in fact, irreversible. Once latency has been established, this copious and silent reserve of viral genes will be useful (to the virus) in a more advanced stage of the infection.

However, the virus does not remain latent in all T helper lymphocytes. It reproduces at a very high speed in many cells, exploiting the host's machinery for the synthesis of new viral particles, to its advantage. During the incessant replications of HIV's genetic material, mutations occur at a frequency rarely observed with other viruses. They appear mainly when RNA—the virus' genetic material—is replicated.

The continuous, rapid evolution of HIV, with its nonstop replication and new mutations, means that the infected human organism is populated not by a single type of virus, but by a family of related, yet distinct, viral particles. As a result, whenever the B lymphocytes produce an adequate number of protective antibodies against a certain type of HIV antigens, new viral particles are already present, with new and different antigens. In other words, it is as if the immune system's productive and reactive capacity is unable to keep up with the whirl-wind rhythm of the virus' constant "changes of clothes."

## Attempts to make an HIV vaccine

Despite its continuous genetic mutations, something exists in HIV that does remain constant. It is called gp120 and it is a protein component of the viral envelope, essential for the HIV genome's entry into helper T lymphocytes. In fact, the protein binds to a lymphocyte surface molecule, called CD4, and uses it as a dock to introduce HIV into these cells.

The gp120 protein was considered the most logical antigen to focus on for developing a vaccine, and an obvious place to start from the perspective of traditional vaccinology. It is a well-known protein and plays a crucial function for the proliferation of the virus. Yet, the vaccines produced against gp120 (using the same recombinant approach that was successful against the hepatitis B virus) failed to live up to expectations. *In vitro*, antibodies against gp120 only neutralize the virus from which the protein is produced, but remains ineffective against the myriad of other HIV particles with which humans can be infected after vaccination. Clinical trials of vaccines of this kind have been conducted with the AIDSVax vaccine by the VaxGen company based in San Francisco, and by the French company Sanofi Pasteur, with the vCP1521 preparation, a modified version of the same vaccine.

This failed attempt led to the discovery that gp120 is protected from an attack by the immune system thanks to a thick coating of sugars, and a distinguishing three-dimensional conformation, so that only the less constant and recognizable parts of the protein are exposed.

A completely different approach to an HIV vaccine arose from observations made *in vitro* and in laboratory animals on cytotoxic T lymphocytes, the cells of the immune system that suppress virus-infected cells. These observations revealed that cytotoxic T lymphocytes were able to kill part of the cells infected with HIV (although not the cells in which the virus was latent).

The researchers reasoned that a vaccine capable of stimulating the cytotoxic T lymphocytes could at least limit the viral reproduction and the progression of the disease, even if it could not prevent the infection.

An attempt was made to build a vaccine that would stimulate cytotoxic T lymphocytes to react against HIV. Unfortunately, even this endeavor was unsuccessful: in two clinical trials on human volunteers, a vaccine with an unpronounceable name (MRKAd5 HIV-1 gag/pol/nef), tested by Merck laboratories, failed to protect people from infection and to reduce the viral load in people infected with HIV after inoculation. What are the reasons for this second failure? Maybe the T cells were not the right type? Was there an insufficient quantity of induced cells? Unfortunately, we do not know.

Short terminology break: the *viral load* is a measure of the severity of an infection caused by a virus; it can be estimated with techniques such as RT-PCR that measure the amount of genetic material of a virus in a sample of mucosa or body fluid involved in the infection.

Another clinical study, conducted in Thailand, evaluated the combination of two different vaccines (ALVAC HIV-vCP1521 and AIDSVAX B/E-gp120). The approach chosen is called *prime-boost*: two different vaccines are administered sequentially, with the aim of inducing a stronger and more complete immune response. The first vaccine consisted of a viral vector, obtained from an inactivated avian virus, in which the modified versions of three HIV genes (env, gag, and pro) were inserted, while the second was composed of the genetically modified gp120 protein.

Vectors are generally weakened or inactivated viruses, completely harmless to human beings, and used as containers and vehicles for the inserted vaccines. Once injected into the body, the vector is like a vehicle that brings the vaccine in contact with the immune system's cells and antibodies. This specific vector, called ALVAC, is an inactivated form of canarypox, a virus that infects birds, does not grow in human beings, and does not cause disease in humans.

The efficacy of this combination of vaccines in the clinical trial was approximately 31 percent—a level of protection considered too modest to bring the vaccine to market. The Thai study was therefore terminated in 2009. However, even a limited result like this could make the difference between life and death for millions of people, especially in a country like South Africa, where 7.5 million people are currently infected with HIV. Researchers therefore tried to improve the two vaccines used in the Thai study, by starting another clinical trial in South Africa in 2016. Unfortunately, as no efficacy was observed, the South African study was also terminated in early 2020.

Among the countless other preparations produced in recent years, the vaccines we have talked about are so far the only ones to have reached phase III clinical trials. And this is despite huge investments, thousands of researchers, and many laboratories committed to this very tough challenge.

Phase III is the last and largest phase in vaccine clinical testing, involving at least ten thousand people. Its aim is to verify that the safety and efficacy observed in the earlier stages are confirmed, as well as identify any rare adverse reactions. The earlier phases are phase I, where a vaccine is injected for the first time into human beings, usually in a group of a few dozen people, to evaluate its safety, dosage, and possible side effects; and phase II, where the study is expanded to a larger group, usually of a few hundred people.

Sadly, repeated HIV vaccine failures have alienated many talented scientists from this area of research, given the number of seemingly insurmountable obstacles that are deeply frustrating.

Despite all this, obtaining an HIV vaccine remains a top priority, particularly as over 4,000 new infections are added every day to the more than 37 million people currently living with HIV.

In fact, there are reasons for optimism. Among those who are HIV-positive, some very rare individuals produce neutralizing antibodies that are reactive against most HIV strains. Another subgroup of infected people has managed to keep the HIV viral load below the detection limits of the most sensitive tests on the market without the use of therapies, in some cases even for 20–30 years.

Unfortunately, we do not yet know why these individuals' immune systems are able to do something that does not happen in millions of other HIV-positive patients. When we will understand how to induce their bodies to imitate those lucky few, perhaps we will have a more promising vaccination strategy than those attempted so far.

An effective HIV vaccine is likely to induce both cells capable of producing neutralizing antibodies and cytotoxic T lymphocytes capable of killing infected cells. The vaccine will also need to recognize most HIV strains and reach the site of infection in record time before the virus gains latency.

Will it be possible to create a vaccine that can do all of this? Perhaps, with a more exhaustive search of conserved antigens common to the multiple HIV genomes, alongside structural studies to observe, for example, how immune cells and antibodies in rare resistant individuals act when they come into contact with HIV antigens. But even this effort may not be enough if we do not initially come to understand the many things that still elude us about the behavior of the virus, and its interaction with our organism.

## A vaccine against malaria: Is it possible?

Every minute, two children die around the world from a parasite of the genus *Plasmodium*, transmitted by an *Anopheles* mosquito. WHO data indicate that roughly half of the world's population—over 3.9 billion people—live in countries of Asia, Africa, Central and South America, where the disease is endemic.

Malaria is not an emerging disease: cyclic fevers like those caused by this small parasite were observed 2,000 years ago by the Greek physician Hippocrates and are even described in 4,000-year-old Egyptian papyri.

Italy had a long history of malaria, so much so that the international name of the disease has Italian origins. *Mal'aria*, literally meaning "bad air," reflects a time when it was believed that the disease was caused by the marshy regions' unhealthy air. The most affected areas were in fact close to ponds and marshes, as well as the plains formed by the riverbeds, near the estuaries. We now know that the larvae of *Anopheles* mosquitoes, the essential vectors for the spread of the disease, thrive in those habitats.

In Italy and other countries, malaria epidemics began to subside when the land reclamation campaigns, initiated in the first half of the last century, were completed. Since the end of the Second World War, the use of drugs like chloroquine, and insecticides such as dichlorodiphenyltrichloroethane, was also crucial in eradicating the disease.

Malaria is still widespread in those temperate and tropical regions characterized by high humidity, stagnant waters, and mild temperatures: ideal conditions for *Anopheles*. Unfortunately, the land reclamation campaigns that make life difficult for the *Anopheles* mosquito cannot be carried out in territories this large. Here, drugs and insecticides have been the main weapons against malaria, but today they are somewhat blunted because of the resistance developed by both *Plasmodium* and *Anopheles*.

Four species in the genus *Plasmodium* infect humans: *P. falciparum*, very virulent and responsible for the greatest number of deaths, especially in Africa; *P. vivax*, the most widespread but less aggressive species; *P. malariæ*, not very widespread and capable of persisting in the body for decades without any symptoms; and *P. ovale*, rare, and occurring almost exclusively in West Africa.

*Plasmodium* spends its life passing from mosquito to human and back. In insects it reproduces sexually, while in humans it multiplies asexually—first in liver cells, and then repeatedly in red blood cells.

The first symptoms of malaria—fever, chills, exhaustion—appear between ten and sixteen days after infection, when the parasites collectively break the membrane of red blood cells to spread and infect other cells. The cycle is only interrupted if drugs are administered or if the immune system somehow manages to take over. Otherwise, the consequences can be coma and death.

Survivors of malaria, if they contracted it as children, usually have built a certain level of immunity to the disease.

Why is it so difficult to get a malaria vaccine?

*Plasmodium* is not a bacterium or a virus, but a protozoan: although it is made of a single cell, it is an organism of far greater complexity than the simpler pathogens. Moreover, it is a parasite with a life cycle that—as we have seen—is very complicated.

Complex and adaptable, the *Plasmodium* has survived for millennia assuming multiple forms and colonizing different species, continuously negotiating a cumbersome cohabitation both with the human organism and with a mosquito's minuscule body.

This long cohabitation has also stimulated human beings to defend themselves. We said earlier that numerous infected children become resistant to the disease—good news, also as proof of principle that the immune system can react to *Plasmodium*. Unfortunately, we do not know why some people become immune and others get sick. Most likely, it has to do with genetic variables that have allowed some humans to adapt to life alongside *Plasmodium*.

Many other animals besides human beings become ill with malaria, but no species seems capable of developing the kind of natural immunity that has been observed in our species. This means that no laboratory animal can help us evaluate the efficacy of candidate vaccines before clinical trials. Only tests in human volunteers, in areas where malaria is endemic, can really show us if a vaccine is effective or not. This greatly increases the difficulty of developing these preparations.

In short, the *Plasmodium* is an adversary worthy of our respect, and so it is not surprising that a truly effective vaccine is still missing.

## Attempts to make vaccines against malaria

At the time of writing, there are almost thirty preparations in various stages of experimentation. Each vaccine falls within one of three development strategies, which mirror the three main life stages of the *Plasmodium* life cycle.

- Vaccines for the *pre-red blood cell phase*: this is the phase in which the parasite enters the bloodstream of an individual bitten by a mosquito and travels to the liver cells, where it will mature and begin to proliferate. Vaccines designed to block *Plasmodium* at this stage should prevent infection or attack infected liver cells.

- *Blood phase* vaccines: this is when plasmodium intensively replicates in red blood cells and is at its most destructive. The vaccines designed for this phase would be unable to stop the infection, which has already occurred, but they should reduce the number of parasites in the blood.

- Vaccines for the *transmission phase*: this is the phase in which the parasite passes from humans to mosquitoes. Here the vaccines should prevent *Plasmodium* from maturing in the insect, after it has sucked the blood of a vaccinated person. These would be vaccines that would be unable to prevent infections or symptoms, but should limit the spread of the disease.

Only one vaccine—called RTS,S/AS01E—has passed all three phases of clinical trials and received a favorable assessment from the European Medicines Agency (EMA) in 2015. Stemming from a military interest (malaria had severely tested the US Marine forces deployed in Vietnam), research for the vaccine began in the 1980s in Maryland, at the Walter Reed Army Institute of Research and the NIH. After a final and lengthy trial carried out in sub-Saharan Africa, the vaccine has proven to be safe and effective in reducing malaria in African children. Today, it is manufactured by the pharmaceutical company GlaxoSmithKline (GSK) alongside an implementation program presently underway for use in three African countries, and is expected to be completed by 2022. If the results prove favorable, the use of RTS,S/AS01E could also be extended to other African countries. The WHO recently recommended this vaccine.

The vaccine is based on the protein circumsporozoite (CSP), which is a molecule involved in the initial phase of infection, before reproduction in red blood cells. It is a recombinant preparation that fuses a part of the CSP protein with the hepatitis B antigen. The idea is that the two protein portions, acting together, induce a more robust immune response.

Like most other candidate antigens for a vaccine, CSP was selected empirically, based on partial knowledge of the *Plasmodium* life cycle. RTS,S offers partial (30–50 percent), transient protection against infection in children. However, given the huge numbers of children affected by malaria in areas where the disease is endemic, even a limited protection can save many lives.

It is possible that a more effective protection could be obtained from the inclusion of different preparations, based on several molecules involved in all three phases of the plasmodium cycle. It is not impossible that the vaccine that will finally resolve the problem of malaria will be built on completely different premises.

Since 2002, when the first genomic sequences of *P. falciparum* and *Anopheles gambiæ* were completed, genetic and molecular knowledge has greatly expanded, with the subsequent reading of numerous versions of genomes and the progressive discovery of the functions for most genes. Other sequences, such as that of *P. vivax*, the second causative agent of human malaria, have been completed. There is hope that the analysis of these mines of information, including the human genome sequences, will provide a better understanding of the specifics behind *Plasmodium*'s interaction with its hosts. Molecules that are still unknown, with crucial functions for the life cycle of all *Plasmodium* strains, could emerge from this research work. It is molecules such as these that could allow for malaria to be eliminated forever in the future.

At present, the blood phase vaccines studied so far have not proven effective in humans, although vaccines that block transmission, with trials underway, could be more promising. In addition, the first generation of vaccines against placental malaria have entered the clinical trial stage. Placental malaria is a form of the disease that causes serious problems in pregnancy, with risks of abortions and complications for both mother and fetus.

Finding more effective vaccines is indispensable, more so now than ever before. The number of victims from malaria is rising, due to the increasing resistance of *Plasmodium* to drugs and mosquitoes to insecticides. Another very current threat is climate change, which is extending the areas of the planet where *Anopheles* thrive. Other people on earth could, therefore, find themselves exposed to malaria. The vaccines currently in preclinical and clinical trials, and the development of new ones, could receive a boost in the near future from the possible discoveries of new antigens, from studies with human mono-clonal antibodies and structural vaccinology, and from other innovative technologies.

## Tuberculosis: It is time to develop a new vaccine

For a while it seemed like a disease you would only read about in novels, but "consumption" sadly made its comeback, and in 2014 it surpassed AIDS for the global number of deaths. According to WHO statistics, every five minutes *Mycobacterium tuberculosis* kills about fifteen people and infects ninety-five.

TB is spread through the air by coughing or sneezing. Once the bacteria have entered the body, they generally reach the lungs where they are recognized by macrophages, the complement, and other molecular sentinels ready to intercept them. Mycobacteria end up being eaten, or more technically, *phagocytized*, by macrophages, which enclose them in an internal compartment, the *endosome*.

At this point, the macrophages, gorged with mycobacteria, migrate to the nearest lymph nodes and display their antigens on their cell surface. The exposure attracts T lymphocytes, which launch a deadly attack against the infected cells, with molecules acting as "perforating hammers" and "grenades."

Despite the violence of the attack, infected macrophages survive in about 30 percent of cases. The resistance is due to the captive bacteria, which are able to inhibit the immune response from within their endosomal prison. For instance, they can change the pH inside the endosome or neutralize the toxic molecules released by T lymphocytes.

Faced with such a stubborn enemy, T lymphocytes accumulate around the infected macrophages and form a spherical cluster, called a *granuloma*. A granuloma is the physical sign of a truce between host and invader. The immune system is unable to annihilate the mycobacteria together with the infected cells, unless it triggers an even more violent reaction, but this would seriously harm the host tissues as well. To avoid this level of collateral damage, mycobacteria are kept at bay by a chronic inflammatory reaction, depriving the invader, barricaded in its fort, of oxygen and nutrients.

As long as this balance holds, TB remains latent. In 90 percent of cases, the latency lasts as long as the infected person's lifespan, but in others the disease manifests itself at some stage.

*M. tuberculosis* can be kept in a state of siege as long as a sufficient number of T helper lymphocytes are available to deal with it and the defense is adequate. But if the defenses weaken, the mycobacteria take advantage of this opportunity and regain their strength. This happens especially in the elderly, in individuals suffering from diseases that weaken the immune system, such as AIDS, in people particularly susceptible to *M. tuberculosis*, or in the presence of some very aggressive bacterial strains.

The proportion of infected people who become symptomatic (5–10 out of 100) might seem negligible, if a quarter of the world's population—about 1.9 billion people!—did not already live with mycobacteria. This is an immense human reservoir, with *M. tuberculosis* awakening at some point in the lives of about 100–200 million people.

The disease is infectious only when it is manifest. Active TB comes in many forms, depending on the organs it affects. In children, the riskiest infection is tuberculous meningitis, while the more prevalent form among adults affects the lungs.

Given that HIV-induced immunosuppression favors TB, it is not surprising that the incidence of TB is growing strongly in all regions where AIDS is endemic or widespread: in Africa, Asia, in Eastern European countries, and in the former Soviet Union.

TB was once treated with antibiotics. However, today most strains are resistant to almost all commonly used drugs, and in some cases even to the most powerful drugs, generally administered in hospitals.

## Attempts to fight TB with vaccines

A vaccine against TB exists, and it is almost a century old. Bacillus Calmette-Guérin (BCG) is prepared entirely with bovine TB bacteria (*Mycobacterium bovis*), live and attenuated. The attenuation was achieved by starving the mycobacteria for thirteen years, through subsequent passages in progressively less nutritious culture media.

Safe and cheap, BCG has also been the most widely used vaccine in the world before COVID-19 vaccines. Every year it is administered to around 100 million infants and children, in countries where the cerebral form of the disease is widespread. Yet over 1.5 million people continue to die of TB each year. What is wrong with this approach?

BCG effectiveness is extremely variable from region to region and from population to population, with protection rates ranging from around 80 percent in the United Kingdom to 0 percent in Malawi, Africa.

Since 1921, versions of BCG that have been grown in many labs around the world have diversified from one another: the strains have evolved from the original vaccine, along with their ability to protect against TB.

But even where the vaccine has maintained its effectiveness, the protection it can offer is limited to about 75 percent against all forms of TB affecting the meninges in very young children. Also, the vaccine protects infants from death, but it does not prevent infection, and it is not effective against the pulmonary form in adults.

Some theories have been suggested as to why protection diminishes over time. The environment in which we live is teeming with germs belonging to the same family of mycobacteria, which includes *M. bovis* and *M. tuberculosis*. We know that some related species of mycobacteria have antigens in common with *M. bovis* (after all, when the BCG vaccine works, it is precisely because *M. bovis* also has antigens in common with *M. tuberculosis*). If a person has been exposed to any of these related mycobacteria before receiving BCG, it is possible that they have already developed antibodies that can at least partially neutralize the vaccine.

Also, the genetic differences between the many strains of both *M. bovis* and *M. tuberculosis* are probably responsible for the limited effectiveness of the old vaccine.

These theories are all plausible, but the reality is also that, almost a century after its introduction, we still do not know exactly why the BCG vaccine provides protection in some cases but not in others.

A new vaccine is urgently needed, but its development is not going to be easy, for several reasons.

First, most TB cases occur in individuals that have effectively controlled an initial infection, but then fail to generate a protective immune reaction once the *Mycobacterium* reactivates. In other words, the immune system against TB works a little (good news), but not completely (bad news), and we do not really know why (terrible news).

Other complications to keep in mind: about a quarter of the world's population is already infected in a latent form; the infection is often accompanied by the concomitant presence of HIV; and most of the people exposed have already been vaccinated with BCG.

What should an ideal TB vaccine do? It should protect against all forms of TB, at every stage of infection, whatever the age or state of health of the population. It should not interact with BCG immunity which is present in millions of vaccinated people. It should act both at the beginning of the infection, preventing the *Mycobacterium* from barricading itself in the endosomes, and in the late phase, blocking reactivation or reinfection. It should work both in individuals who have never encountered *Mycobacterium* and in people who have already been exposed.

Are these too many things to ask of a vaccine? Possibly. Perhaps this is why research is aiming at a number of immunizing products, each of which may be able to solve at least part of the problems we have listed, instead of focusing on a single preparation.

For example, live and attenuated recombinant preparations are being developed, starting from BCG, to improve the existing vaccine by extending coverage to pulmonary TB, or by blocking the *Mycobacterium*'s ability to hide inside the endosome. However, vaccinologists must be very cautious because all attenuated vaccines carry the risk of mutations and reactivation, a very serious problem in immunocompromised people, such as those living with HIV.

Vaccines are also being developed to work as a boost to the immune response for those people who have already been infected in the past, in the event of reinfection and reactivation of the mycobacteria.

Today, a number of vaccines against TB, designed with a variety of formulations, are in different stages of development and clinical studies. There is some hope, especially for the M72/AS01E vaccine, which at the end of 2018 was shown to provide approximately 54 percent protection against the disease in a phase II clinical trial conducted in Kenya, South Africa, and Zambia. The study involved 3,300 individuals with latent TB, i.e., the silent infection that becomes manifest in about 1 out of 10 cases, usually within the first 2 years of infection. The subunit vaccine, designed to provide protection against active TB, is derived from the fusion of two *M. tuberculosis* antigens (32A and 39A), to which an adjuvant (AS01E) has been added. The 50 percent success rate achieved so far may seem modest. Still, even a partially effective vaccine could save millions of lives, if the results are confirmed in larger and longer studies, given the mind-blowing numbers of TB—10 million new infections and 1.5 million deaths per year.

## Hepatitis C: Why drugs are not enough

Hepatitis C is a liver disease caused by hepatitis C virus (HCV). It is a single-stranded RNA virus, which in some cases causes a mild and short-lived disease, but in about half of the infected people gives rise to a serious and permanent disease. Worldwide, 58 million people live with chronic hepatitis C and a significant number of those infected will develop cirrhosis or liver cancer during their lifetime. It is these secondary diseases from HCV that cause the deaths of many hundreds of thousands of people each year: 290,000 in 2019 only.

HCV transmission occurs mostly through blood, for instance, with the exchange of syringes among drug users, or through sexual practices that involve the exposure to traces of blood. In the past, when it was difficult to test donor blood and its derivatives for the presence of the virus, many people were infected from blood transfusions.

There are antiviral drugs which theoretically can cure many people infected with HCV, reducing the risk of death from cirrhosis and liver cancer. But these treatments have important side effects, they are not always well tolerated by patients, and their rather high cost limits their access to a small number of patients.

To date, we have no effective vaccine against hepatitis C, which means that prevention is entrusted to the individual, who should avoid taking risks, and to the health sector, to maintain high standards of hygiene, and to always test donor blood and its derivatives for the virus.

Attempts to develop a vaccine began about thirty years ago, when Harvey Alter, Michael Houghton and Charles Rice finally managed to identify and grow the hepatitis C virus, after countless attempts to hunt for a very elusive microbe. They were awarded the 2020 Nobel Prize in Physiology or Medicine for their discovery.

More than twenty potential vaccines have been studied in the lab, and some of these have entered initial clinical trials.

Progress has been slow for a number of reasons, including the fact that the virus possesses some unique characteristics: the disease is caused by at least seven genetically distinct viral variants, each in turn characterized by multiple subtypes. In total, at least fifty subtypes have already been identified. An effective vaccine should at least protect against those variants that cause most of the infections around the world.

Also, until a few years ago, it had not been possible to grow the virus in cell culture, and today still, no one has succeeded in replicating the human infection in small laboratory animals such as mice. The HCV can, in fact, infect chimpanzees, and the infection in these animals is similar to that of adult humans. Research in primates is limited for ethical and financial reasons, but animal studies remain fundamental and unavoidable for the development of a vaccine.

However, there are grounds for optimism. In a significant number of people with acute HCV infections, the virus is eliminated spontaneously, with a specific immune response. This is a proof of principle that the immune system can recognize and eliminate the virus: a good sign for the development of a vaccine.

There are still several obstacles that will need to be overcome. Firstly, it is not yet clear what effects the vaccine must produce on the immune system in order to activate a protective effect, nor the nature of the effect: against the initial infection or against the onset of a chronic infection? In addition, a suitable population at risk has to be identified for carrying out clinical trials. That is not an easy task, especially in Western countries, where people have been educated to avoid behavior that facilitates contracting HCV. Finally, we need more research into the protein structure of the virus against which a vaccine could be developed, to understand what has gone wrong with the candidate vaccines developed so far. Electron microscopy studies have actually shown that the recombinant E2 protein, used in some experimental vaccines, is much more flexible in the conformations it assumes than its natural counterpart, which instead appears to be more rigid. This could be one of the reasons why the immune system struggles to produce effective neutralizing antibodies against the HCV virus.

## No hope, high hopes?

By this stage you would not be wrong if you believed that a very long, uncertain, and bumpy road lies ahead in the search for vaccines against HIV, malaria, TB, and hepatitis C.

With malaria and TB, the outlook is more promising than in 2009, when we wrote the first edition of this book.

As for HIV and HCV, we know that a purely empirical approach should be abandoned in favor of a more exhaustive and rational analysis of every possible antigen, starting from genomic and structural biology studies. But this methodological change, while necessary, may still not be enough to identify the most promising vaccine candidates. In other words, it is probably time for a more pronounced "paradigm shift."

The problems that remain to be solved are immense, but it is also true that their solution may be just around the corner. Remember how impossible the challenge for the Men B vaccine had seemed? Before reverse vaccinology opened up a path, no one could imagine how to get out of the "tunnel of impossibility," and the likelihood of an anti-Men B vaccine was zero. Yet we made it, against all odds, and within a very short time.

How will the journey of ideas proceed? Who will win this marathon run of attempts? Will these seemingly elusive vaccines ever see the light of day? The evolution of discoveries is unpredictable, but that is also the beauty of science.

# 6
# Unpredictable Viruses: The Flu

Before COVID-19, the most prominent disease that could suddenly appear and travel the globe was undoubtedly the flu. Each year, over the winter months, one or more flu viruses materialize, varying slightly from previous seasons. This variation reflects a deeper change: a constant genetic reshuffling that, in an infected individual, can happen every six hours, when an arsenal of new viruses emerges from an infected cell, ready to conquer other cells and individuals.

Continuous mutation is an excellent evolutionary strategy for the virus: the immune system is forever unprepared to recognize old acquaintances that show up with brand-new features—theoretically it is possible to get sick with flu every year. To this, add the fact that different flu viruses can jump relatively easily from species to species: the flu infects pigs, birds, humans, horses, seals, cats, dogs, tigers, ferrets, etc.

Not that every flu virus can infect all these different animals. Fortunately for us, each species has its own pool of flu strains, geared by preference towards infecting only those animals and not others. But given the extreme genetic variability of this virus family, jumps between species are frequent, with an almost limitless animal reservoir exacerbated by human activities. We raise massive quantities of animals for our own consumption, while at the same time carelessly modifying the habitats of many wild species. All this creates endless opportunities for encounters between us and other animals, and our respective viruses.

Human beings often get infected with flu viruses from pigs and birds in places like farms, fairs, and live animal markets. However, avian and swine flu viruses usually do not reproduce particularly well in our bodies. But what happens if an avian virus and a human virus end up in the same cell together? A *superinfection* occurs, where the genes of the two viral strains may cross.

## What do we know about flu viruses?

Three types of influenza viruses—A, B, and C—are known to infect humans. Types A and C are promiscuous, that is, they can infect multiple species, while type B almost exclusively affects us alone.

The genome of the influenza virus is made up of eight segments of RNA: six of these are used by the virus to make copies of itself, while the other two help the virus build two proteins that are crucial for infection and dissemination.

The two proteins, called hemagglutinin (HA) and neuraminidase (NA), are both found on the outer surface of the virus. HA allows the viral particles to penetrate into the cells of the host's respiratory tract, where the virus then multiplies by exploiting the host's resources. NA, on the other hand, allows newly assembled viruses to leave an infected cell and reach other individuals (through the now famous droplets spread by breathing, coughing, and sneezing, which have been crucial to the COVID-19 contagion as well).

Given the frequent mutations, these proteins can take many different forms: we know at least eighteen HA and eleven NA proteins. Each influenza strain is identified by the combination of specific types of NA and HA, which are also found in its identification code, abbreviated as HnNn (H and N are the initials of the proteins, while n is the number identifying the type).

For instance, the pandemic known as the "Spanish" flu, which may have killed more than forty million people between 1918 and 1919, and the one that caused the pandemic popularly known as "swine" flu, in 2009, were both caused by a type A H1N1 virus. On the other hand, the 2003 avian flu, which was almost exclusively infecting birds, was caused by a type A H5N1 virus. Subtypes H1, H2, H3, and N1 and N2 are the most prevalent traces found in the immune memory of the human population.

Interestingly, of the 18 known HA subtypes, only 6 infect pigs and 8 humans, while all but 2 are found in birds. Since similar ratios apply to the known types of NA, it is clear that a bird's body was the most convenient way for a flu virus to travel around the world, prior to the invention of the airplane.

## The seasonal flu

Influenza, like other viruses of the respiratory tract, causes annual epidemics, and in temperate regions can peak in the winter season. What flu viruses do during the summer, and why they follow a seasonal cycle, is one of the many puzzles that make this pathogen fascinating (and worrying) for its observers.

Statistics for the flu's global disruption each year estimate around 3–5 million serious cases and up to 650,000 deaths (with roughly 11,000 deaths in the United Kingdom). These numbers are far from negligible, with victims especially among the elderly, the chronically ill, and the very young.

However, you can protect yourself against seasonal flu. For quite some time now, a vaccine is brought out at the beginning of autumn, specifically produced every year to protect against the main kinds of flu circulating that season. How is this vaccine made? Let us take a look at the production steps that are repeated yearly.

## Surveillance

Worldwide, 148 institutions exist in 124 countries, whose job is tracing every flu strain circulating in each territory or country. They are called "national influenza centers" and they make up the WHO global influenza surveillance network. This surveillance activity, already extensive in the past, was further strengthened after the 2003 and 2006 avian flu outbreaks, and with the 2009 H1N1 pandemic. These three outbreaks were warning signs alerting the world to the ever-present risks of flu viruses.

The Internet is a particularly effective information tool for this kind of activity, as it allows for real time communication of the strains, old and new, that are becoming prevalent at any given moment in every corner of the planet. Also, the Internet and social networks can provide some very useful clues for tracking the flu: when searches for symptoms such as "sore throat," "fever," and "flu" intensify over a certain period, in a certain geographical region, an outbreak may be occurring. These indirect but important signals are closely monitored by the surveillance network.

## Isolating the virus

In practice, how do national influenza centers find and isolate new viruses? Well, have you ever been sick with flu and your doctor has taken a nasal or throat swab, like the ones used to test for COVID-19?

Essentially, by taking a mucosa sample of the inside of the nose and the back of the throat from a person with symptoms, using a long cotton swab, the material collected can then be subjected to analysis by RT-PCR, a technique that has become famous during the coronavirus pandemic. With this method, the quantity of genetic material present in the sample is multiplied many times, making it possible to verify the presence of gene sequences belonging to a known, or as yet unknown, microorganism.

An older and slower method can be completed in a lab by growing the microorganisms that have attached themselves to the swab in cultured cells. Assuming that the virus replicates in cell culture—some viruses are quite reluctant to do this!—both its replication rate and its ability to infect new cells can offer useful information on the dangers posed by the virus.

If an influenza virus is found from either method, it will be isolated from the other germs. The sequenced genome of the new virus will then be characterized and compared to those for all other known influenza viruses, to establish whether we are dealing with an old acquaintance or with a new result of evolution.

Today, we know how to reconstruct the evolutionary history of many influenza viruses in detail, thanks to these kinds of analyses and to the collection of data in large databases, accessible to all researchers in every corner of the world.

## The terrible four

Flu vaccinologists are, to a certain extent, like fashion designers, "sniffing" for vibes that are trending for next season's clothes designs. Every year between November and January, WHO experts try to understand which flu viruses are the most worrisome among those identified by the surveillance network. Usually, these viruses first appear in Australia, Asia, sometimes in America, and rarely in Europe. A selection is made of the ones most likely to spread in the next flu season.

Fortunately for us all, the WHO forecasts rely on more dependable tools than those used in fashion. With the help of computer models, the forecasts rely on past experience, on the analysis of the genetic characteristics for each virus, and on predictions for its future behavior.

Experts usually choose four of the identified viruses to develop a vaccine for each seasonal flu. Once the four strains have been chosen, the countdown starts for the manufacturers to produce the new vaccine.

## From the viral genome sequence to the synthetic virus

Until the first decade of this century, most isolated flu viruses were physically shipped to a WHO laboratory, which could be in Australia or London, or to a CDC lab in the United States. Here, the viruses were adapted to grow in eggs, before being physically shipped to companies around the globe for vaccine manufacturing. Many viral journeys were far from ideal, both for the time taken to ship across borders, and for safety reasons.

A major change occurred in 2013, when a potentially pandemic H7N9 flu virus began circulating in China, infecting three people and killing two. It was an Easter Sunday and the Chinese health authorities had published the sequence of the HA and NA viral genes online. That same day, the team of Rino Rappuoli, one of the authors of this book, contacted a collaborator of Craig Venter, who was in San Diego, California.

Venter immediately synthesized the two genes, starting from their gene sequences. On Monday evening, he shipped the two synthetic genes from

San Diego to Rappuoli's lab in Cambridge, Massachusetts, where they arrived on Tuesday morning. The synthetic genes were immediately used to make two types of vaccines. One was a fully synthetic RNA vaccine which, in less than a week, was ready for testing in laboratory animals. The other was a conventional vaccine produced by growing a recombinant influenza virus generated starting from the genetic sequence published online. In addition, this candidate vaccine was ready in less than a week for manufacturing and subsequent evaluation in clinical trials.

Using standard, traditional procedures, the US CDC labs were able to send the virus to vaccine manufacturers much later, a few months after the one made with *synthetic vaccinology*, as this new process was called.

Fortunately, the H7N9 virus was not able to transmit effectively between humans, and the scare quickly evaporated. But what had begun merely as a proof of principle—to verify if it is possible to build a vaccine starting from the shared information on its genes and not from a physical virus—was, in fact, a turning point. Since then, viruses do not need to be shipped any longer, and the dangers associated with this method are now a thing of the past. Only the information about their genetic sequences is shared via the Internet, and the viruses can now be artificially synthesized in labs around the world to facilitate the production of vaccines. This new procedure, that was later named *Internet-based vaccines*, was a totally new way of producing vaccines back in 2013. It exploits the Internet to instantly share the necessary viral genomic information with any laboratory around the planet, avoiding the hazards of transporting viruses around the world for delivery to a limited number of laboratories. This method obviously works with any kind of virus, not just flu, as we were able to see in real time, in the case of SARS-CoV-2 in January 2020.

## From the synthetic virus to the master seed

Once the viruses from the four selected strains have been synthesized, it is necessary to grow them in very large quantities, should we need the entire virus and not just the individual proteins to produce the seasonal flu vaccine.

But viruses can replicate inefficiently or unevenly in the lab, which can be an obstacle to mass industrial production of a vaccine.

To solve this problem, manufacturers try to "normalize" a virus' growth rate using the following procedure:

- Cells susceptible to the virus are infected simultaneously with each of the four new strains and with an old, vigorously growing strain (PR8).

- While the viruses reproduce inside the cells, the PR8 and the new strains exchange fragments of RNA. In this way they give rise to a progeny of viral particles, called *reassortants*, that have a mixed genome with varying amounts of genes from both old and new strains.

- From this mixture, the researchers select the viruses that have the two genes to build NA and HA, originating from one of the new strains, and the rest belonging to PR8.

- This procedure allows for the construction of a hybrid virus which has the "engine" of PR8, a strain capable of reproducing itself in a very efficient way, and the "bodywork" of the new strain (composed of NA and HA).

The procedure is then repeated for the remaining three strains.

The next step is to obtain a larger quantity of virus, the so-called *master seed*: a stock of viral particles, from which all vaccines will be produced.

Those who handle viruses protect themselves from infection with clothing that includes special masks, double gloves, surgeon's caps, and other protective measures, specified in what is called "Good Manufacturing Practice" (GMP), a set of rules that are also necessary to obtain the product's approval.

## Vaccines grown in eggs

For many decades, viruses for vaccine production were grown in fertilized chicken eggs. Before inoculation, each egg was illuminated with a special light to see if the embryo was alive (otherwise the virus would not reproduce). Then the date and time were marked on the shell. Finally, with a drill like the ones found in a hardware store, a hole was made at one end of the egg, and a dose of virus was injected.

If this method seems quite homespun and low-tech to you, you are not wrong. In fact, the use of eggs in the production of vaccines is a technology that was developed almost a century ago, and it survives in old manufacturing plants alongside the more high-tech procedures in place since the start of the genomic era.

To their credit, although a somewhat antiquated method, chicken eggs are still an excellent and cheap way to grow viruses. About 80 percent of seasonal flu vaccines are still produced in this way. This is because eggs are natural fermenters, their *allantoid liquid* containing everything that is necessary for viruses to proliferate vigorously for 48–72 hours. Once the growth is complete, the liquid is aspirated and rotated at high speed in an *ultracentrifuge* to separate out the virus.

The only real drawback to an egg-based process is productive capacity, and consequently, the amount of time required to produce the vaccine. It is estimated that the maximum number of vaccine doses that can be obtained every year in the world with this system are about four hundred million (a number that has undergone some upward adjustments, thanks to investments made in recent years).

## Cellular vaccines

To overcome this constraint, new systems for growing vaccines in mammalian cell cultures have been developed that, unlike eggs, allow for the production of many more doses, and at a faster rate. The only system of its kind that has so far been approved by the FDA in the United States, in 2016, belongs to the Seqirus company, and is used for the production of inactivated influenza vaccines. The system is also in an advanced stage of review for approval at EMA. In the space of a few years, cellular vaccines may send eggs into retirement: a well-deserved rest, after more than ninety years of honorable service.

## Recombinant vaccines

An even faster and more efficient method exists by obtaining the proteins for influenza vaccines from genetically modified cells grown in culture. The cells used in this kind of cell culture can be derived from microorganisms, such as bacteria and yeast; from plants, such as tobacco, or from insects. Also, with this method, only the gene sequences corresponding to the proteins are needed, while the virus itself is not necessary. For all these reasons it is the fastest, safest, and most efficient of the three systems.

Recently a recombinant vaccine produced in insect cells has been licensed. In this case, the gene for HA or NA, synthesized in the lab, was simply inserted into a baculovirus, a virus that infects insects, using genetic engineering. The baculovirus' role is to transport the genetic instructions into a host cell approved by regulatory agencies; in essence it operates like a biotech factory, producing the desired antigen quickly and in large quantities. An additional advantage of cellular and recombinant systems is that its vaccines are tolerated by people allergic to eggs.

## Vaccine purification

Whatever the manufacturing method used, the vaccine must be purified before being inserted into the vials. For vaccines grown in eggs and cells, there are various levels of purification available. In the past, the whole virus was purified and inactivated, but this method is no longer used for influenza vaccines. Today, the vaccines components are split by a detergent added to the preparation, which breaks down the virus parts, but leaves them all together in the vaccine. The most advanced and "clean" method is used for subunit vaccines: once the virus has been broken down into its components by the detergent, the HA and NA proteins are purified and the rest is discarded. In the case of recombinant vaccines, the individual proteins obtained are simply separated and purified from the cells in which they were produced.

## The addition of the adjuvant

An adjuvant can also be added to the vaccine formulation to increase its effectiveness (later in this chapter, we will see what adjuvants are and how they work).

## End of production

Production usually ends between May and July, depending on how many doses are needed. Each vaccine dose is then inserted in a vial, and a series of checks, including sterility tests, are made before shipping the vaccines to pharmacies.

The processes we have just described for the flu vaccine are also used, with minor variations, to produce most modern-day vaccines.

## How effective is the seasonal vaccine?

"I got the vaccine and, damn it, I still got the flu!" Who has not heard this complaint at least once? In many cases, once vaccinated we actually do not get sick from the flu at all and what seems to us to be the flu is actually an infection caused by other bacteria or viruses with similar symptoms. However, it sometimes happens that those who are vaccinated still fall ill with the flu, as the vaccine does not eliminate the risk of infection, but reduces it by about half.

Why is not the coverage higher? Different people, particularly the elderly, can react more or less effectively to vaccines. But the reason lies, above all, in the small mutations that influenza viruses accumulate during the flu season. These changes result in variations between the vaccine strains tested and those that are actually circulating months later, when the vaccines are finally administered. Mostly for this reason, the effectiveness of seasonal vaccines is, on average, good, but not absolute.

Even with these limitations, vaccination against seasonal flu is a very important form of societal protection because it helps prevent the most serious forms of the disease and deaths, while reducing the symptoms and consequences of infections. The flu vaccine also helps contain the number and duration of hospitalizations, preventing the health system from imploding. If we did not at least vaccinate the most vulnerable against the flu, the damage would be much, much greater, comparable to what we experienced with COVID-19, a disease that was not preventable with a vaccine for the entire first year of its appearance.

## Influenza pandemics

Unpredictably, influenza viruses can sometimes appear in more aggressive forms than usual and are sometimes even able to travel around the world. An epidemic that sweeps the planet is called a *pandemic*. However, a globetrotting virus is not necessarily more lethal than one that struggles to travel; we have had examples of both mild and aggressive pandemics.

In just over a century, there have been five influenza pandemics we know of: in 1889, possibly caused by H1; the "Spanish" flu, caused by H1N1 in 1918–1919; the "Asian" flu, from H2N2, in 1957; the "Hong Kong" flu, caused by H3N2, in 1968; and finally, the A flu, H1N1, in 2009.

In all five pandemics, the viruses spread in successive waves, and often, but not always, the second wave did more damage than the first. The "Spanish" flu virus, for instance, appeared in March 1918 in an extremely mild form, but the second wave, in autumn 1918, caused most contagions and deaths, while the third wave, the following winter, was far less severe.

Special combinations of viral genes, combined with particular environmental conditions, are likely to have contributed to the difference in virulence of past epidemics. However, as we do not know the precise reasons why a virus behaves in one way rather than another, it is very difficult for experts to predict what will happen when a new potentially pandemic virus appears. In a dynamic and constantly evolving situation, like that of a rapidly spreading pandemic, every forecast, even the most informed and expert, is inevitably subject to errors, corrections, and revisions. "Making predictions is difficult, especially about the future," said Yogi Berra, a famous baseball player, when talking about sport, but the motto is just as valid to epidemic waves too.

The most recent flu pandemic originated in Mexico. It was caused by an H1N1 virus made of an unprecedented blend of viral genes, from strains that infect pigs, birds, and humans. The new virus originated from at least one spillover event from an animal species into *Homo sapiens*, and the flu it caused burst onto the media on Friday, April 24, 2009, after the WHO declared a public health emergency of international proportion.

This very infectious virus quickly made its way around the five continents, worrying experts because, at least initially, the cases seemed to be quite serious. Vaccinologists and public health experts worked day and night for many months to keep the crisis from flying out of control.

This was the first time that we had the opportunity to follow the spread of an epidemic in real time on a global scale, with extraordinarily fast and efficient communication and surveillance systems. We also had antiviral drugs that could reduce the impact of the disease and contain its spread. And we had a plausible vaccination strategy.

How come we were so well-prepared? To understand why, it is worthwhile going over what happened in the field over the last 25–30 years.

## The fear of the avian flu: An anti-pandemic test

In 1997, experts were alarmed by the spread of the first major avian flu epidemic of our times when a new H5N1 type A virus showed up in Hong Kong. It was a virus that mainly affected birds, rarely humans, except through close contact with infected animals.

The deaths of six people in Hong Kong sparked panic and intense media attention. In the rare cases where the virus succeeded in infecting humans, the disease was very virulent and often lethal. The widespread fear was that one day, the virus would mutate enough to be transmitted from human to human and not only sporadically from animals to man.

Luckily the fear quickly subsided: with the culling of many millions of infected birds ordered by the Hong Kong health authorities, the contagion disappeared rapidly from the news and from the "radar" of the surveillance authorities.

The virus seemed to have disappeared, at least until 2003, when a new epidemic erupted, this time in mainland China. In those six years of latency, the virus had spread silently among farmed chickens and migratory birds, causing an increasing number of human deaths (4 in 2003, 32 in 2004, 3 in 2005, and 79 in 2006).

The somewhat excessive panic that spread through the media to the worldwide population in 2003, and again in 2006, actually had some useful consequences. In the United States, the fear prompted the government to check the country's capacity for vaccine production. After the difficult period for vaccines mentioned in Chapter 4, by 2003 only two or three surviving firms were left in the United States, compared to about thirty in the 1970s. And only one of those companies produced flu vaccines.

The US government decided to invest a significant amount of money—$7 billion—to prepare the country for a possible pandemic, and not just for flu. According to some observers, the initiative was perhaps one of the few positive legacies of the George W. Bush administration, unfortunately weakened later by the very unwise decisions taken by the Trump presidency.

With that funding, significant preparations were made from experimenting with novel vaccines, trying innovative production methods, and carrying out clinical trials on tens of thousands of people. In short, vaccinologists were rehearsing the scenario of a possible human flu pandemic, which many experts believed possible.

The labs in Siena, Italy, at that time part of Novartis, were awarded parts of those funds in light of the international credibility they had acquired in research and development of vaccines. This is an important strategic resource for any country, if you consider that until the COVID-19 pandemic, only nine nations had high-level industrial and research facilities on vaccines, with several advanced countries forced to purchase vaccines abroad. The fact that Siena already disposed of important vaccine technologies, including the adjuvant MF59, also played a part in this decision.

## The adjuvants

Some vaccines are a bit like some people with extraordinary talent and intelligence, but who are a total disaster socially. At parties they end up as wallflowers, with no one paying attention to them. But sometimes, these people have a smarter friend, who by being a kind of showoff, gets them both noticed. Even the best vaccines need help to get noticed by the cells of the immune system—and that is what *adjuvants* are for.

So far, six types of adjuvants are used effectively and safely in human vaccines. The oldest is an aluminum salt (aluminum hydroxide or aluminum phosphate), introduced in 1924 and initially used in diphtheria and tetanus vaccines, after it was discovered that it strengthened the immune response to those preparations. It was later used in other human vaccines throughout the last century.

New adjuvants were later developed, targeting specific components of the immune response, allowing for a stronger and more lasting protection against diseases. These properties are especially useful for the most recent vaccines, as many of the older vaccines are made of whole, weakened, or killed germs, and therefore are alone able to stimulate the body to produce a strong protective immune response. Most modern vaccines, on the other hand, contain only a very small fraction of the components of a germ, sometimes just one or a few proteins from a virus or bacterium. Here, the adjuvants help the body induce an immune response that is strong enough to defend people from the diseases against which they are vaccinated.

In 1997, Chiron was the first company in the world to register a new adjuvant, MF59, now owned by Seqirus. The substance is essentially an aqueous emulsion of an oil nanoparticle the size of approximately a hundred nanometers, or $10^{-7}$ m.

When injected into the muscle along with the antigen, MF59 causes a micro-inflammation capable of attracting all the cells that can introduce the antigen to the immune system. In this way it artificially creates an ideal situation that facilitates the immune response occur in the lymph node (to learn more about the presentation of the antigen and lymph nodes, see the Appendix on our defenses and how they work).

The oil in MF59 is called *squalene*, and it is a biodegradable substance naturally present in most living organisms, from plants to humans. Our body uses it in the biosynthesis of cholesterol, and precisely because of this compatibility with the body it is also widely used in cosmetics. In its industrial production process, squalene is purified from shark liver oil (hence the name). The squalene nanoemulsion is obtained through a micro-fluidization process, a technique developed in the late 1980s.

MF59 is approved in Europe and the United States, is considered very safe, and has been administered, alongside the flu vaccines, to more than fifty million people since 1997. Another adjuvant based on squalene is AS03, owned by GSK.

Many people are afraid of sharks, and few are fond of them. However, since a part of their body has started to be used in vaccines, even this ferocious species has found its own defenders. The concern of some animal rights groups is that too many sharks are being sacrificed to produce adjuvants, particularly for SARS-CoV-2 vaccines, given the massive number of doses that are needed to control the pandemic. However, about 90 percent of the squalene obtained from these animals is used in cosmetics rather than in vaccines.

Possible alternative sources of squalene, both from plant organisms and from synthetic processes, are currently being investigated. Adopting a replacement substance, assuming it can be identified, will still take time before it can be used, given that each adjuvant must be carefully evaluated clinically before receiving approval from regulatory bodies.

There are also other types of adjuvants available in addition to those obtained from shark liver. AS01 is an adjuvant composed of monophosphoryl-lipid A, an immunostimulating substance isolated from the surface of bacteria, while

QS-21 is a natural compound extracted from the Chilean soap tree (*Quillaja saponaria molina*). It is used in a recombinant vaccine, recommended for people over fifty, against *Herpes zoster*, the virus that causes chickenpox and shingles.

Matrix-M is an adjuvant similar to AS01, also derived from the Chilean soap tree, with the addition of cholesterol and phospholipids. It has been approved for use in the COVID-19 vaccine produced by Novavax, which we will be discussing in Chapter 8.

Finally, CPG 1018 is an adjuvant containing a synthetic form of DNA, cytosine triphosphate followed by a guanine, which mimics bacterial and viral genetic material and, in this way, attracts the attention of the immune system. It is used in a hepatitis B vaccine.

All adjuvant-containing vaccines were tested for safety and efficacy in large clinical trials before being authorized for use. Once approved, they are continuously monitored for any rare side effects that did not emerge in clinical studies by regulatory bodies such as the FDA in the United States, the EMA in Europe, and the Medicines and Healthcare Products Regulatory Agency in the United Kingdom.

## Is it possible to beat a flu pandemic on time with a vaccine?

In the H1N1 flu pandemic that began in 2009, experts thought that the best way to protect the population and stop the infection would be through the development and use of a vaccine. This hypothetical "would" indicated that even the most optimistic among them was well aware that a vaccine had never been used before in a pandemic.

Was there enough time to build a vaccine against an emerging, potentially pandemic virus? We have seen that the manufacturing of any seasonal influenza vaccine takes about six months, starting from the moment the new strain is isolated, sequenced, characterized, and made ready for vaccine production.

The genome for the new virus that emerged in Mexico, in 2009, and the method for its molecular diagnosis, were made available very quickly from the very night, in late April, when the news of the epidemic went around the world. The shortening of the time needed for isolating and defining the virus was excellent news: precious time had not been taken away from the vaccine's development and production.

Still, some doubts remained: how many doses of a pandemic influenza vaccine could be produced? And would manufacturers still have the capacity to produce the new vaccine at the same time as the seasonal flu vaccine?

Faced with an unknown virus, it was likely that the population lacked immune memory (the one which, in the case of a virus that is already known, could be awakened by a single vaccine dose). Two protective doses would be needed: a first dose, of *priming*, which would make the new virus known to our defenses, and a *booster* dose, to strengthen the immune memory.

Looking beyond the emergency, an alternative strategy might have been to vaccinate the population with a *pre-pandemic influenza vaccine*, that is, one capable of offering protection against most of the strains considered to be the most dangerous. The vaccine, administered well before a possible pandemic, would have generated lasting memory cells. These could then have been recalled with a single dose of an additional vaccine, in the event of a pandemic.

This approach was more than a mere hypothesis, since a clinical trial had been successfully conducted in 2007 at the University of Leicester, in the United Kingdom. Perhaps such a trial could become the basis for a future anti-pandemic vaccination strategy.

But in practice, what happened in 2009? And what was the balance of the H1N1 flu pandemic? The pandemic lasted about 19 months, from January 2009 to August 2010. The number of people who became infected was calculated between 700 million and 1.4 billion, or between 11 and 21 percent of the world population at the time. These numbers included people who were asymptomatic or who had a very mild form of the disease. The number of estimated deaths caused by H1N1 range from 154,000 to 575,000. In the end, the virus proved to be less lethal than an average seasonal flu, but a more correct comparison is with the mortality rate from earlier pandemics: at least one order of magnitude less than the "Asian" and "Hong Kong" influenza pandemics, and less than two orders of magnitude when compared with the "Spanish" flu. In fact, we were extremely lucky! Especially considering the fact that the vaccine failed to arrive in time.

The vast experience gained with the development and production of seasonal flu vaccines has allowed us to greatly accelerate the race to the anti-H1N1 vaccine, which in 2009 took less than six months, an extraordinarily rapid production time, both in absolute terms and for the time period. This was not enough, however, for the very fast spread of this highly contagious virus. The doses arrived in vaccination centers and pharmacies by mid-October 2009, when the peak of infections had been reached about a month and a half earlier (Figure 1), too late to be of any use. Fortunately, the virus, although very infectious, was not very virulent. In the end it was much less lethal than at the beginning of the pandemic, when many serious cases had been registered, especially among young people.

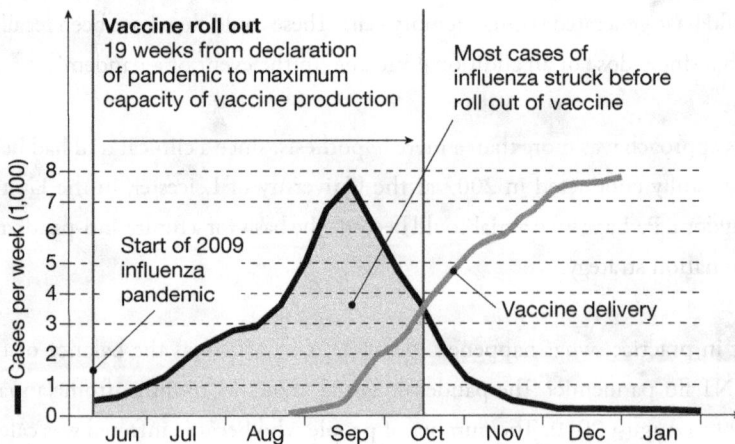

Figure 1: The 2009 H1N1 vaccine was ready in less than six months, a very short time, but not fast enough to beat the virus (adapted from *Nature*, 6/3/2014).

The modest lethality of the virus during its phase of greatest pandemic spread led many to believe that the warning bells sounded by the WHO and by public health authorities in different countries had been greatly exaggerated. In these situations, the duties of officials and experts are ingrate and thankless: they have to provide the public with reliable information about a life-threatening disease, without triggering panic. However, the virus is a moving target, and the situation is always fluid, ever-changing, and unclear. It follows that officials and experts can only make provisional statements and suggest measures that will necessarily be subject to frequent changes. Whatever they say or do, in the public eye they can only do wrong. With hindsight, it is easy to see the 2009 pandemic as an alarm that was fortunately brief and far less frightening than expected, but it could have gone very differently. Still, with hindsight, it is easy to say that millions of people were unnecessarily alerted (or that the underlying reason lay with humongous pharmaceutical and governmental budgets). However, if the flu had actually caused a severe pandemic, with its limited management tools, it is safe to say that officials and experts would have been blamed for their inability to contain it. We have lived both scenarios; in 2009, with the H1N1 flu, and since 2020, with COVID-19.

## Will there ever be a universal flu vaccine?

At this point you will probably have become convinced that, in the evolutionary competition, microbes, and in particular influenza viruses, are way ahead of us, thanks to multiple opportunities for infection in different species.

Victor Vaughan was a physician who, in 1918, was dispatched to a field hospital near Boston to report on the "Spanish" flu that was raging there among young soldiers returning from the war. The virus, he commented, had "demonstrated the inferiority of human inventions in the destruction of human life."

However, we too have made considerable progress, despite the constant challenge from these very swift creatures that are able to adapt with every new opportunity. We have increased our defense capabilities in a way that was unthinkable up until a few years ago, to the point where a flu vaccine covering many strains is now within reach, and a universal vaccine is a possible prospect.

Among the new approaches currently under study, the most promising is focused on HA, the surface receptor of the virus, and in particular on the *stem* region of the protein. This region, shaped like a sapling, is conserved in the numerous strains of influenza viruses, so that a vaccine against it could offer a broad protective coverage against multiple influenza viruses, possibly all of them.

Years of intense work have taught us a lot about influenza viruses, which are among the most monitored and studied microorganisms in the world, perhaps because of the still vivid memory of the decimation produced by the "Spanish" flu. We now really know quite a lot about their genome, the proteins that compose the viruses, their mechanisms of infection and proliferation, and the epidemics they caused.

However, there are still many questions we are unable to answer: why can a certain viral strain pass easily from one individual to another, while another struggles to infect? Why is a vaccine more effective in one person than in another? How can we induce the immune system of the elderly to respond more effectively to vaccination?

To improve the design of our protection strategies, we need to better understand which characteristics of the vaccine, of the virus and of the host's immune system inhibit transmission from individual to individual, and from species to species.

# 7
# A Pandemic of Our Time

Between January 23 and February 2, 2020, about seven thousand workers and over one hundred bulldozers were at work on a construction site in Wuhan, China. Many of us, stunned and incredulous, watched, live, the images of the construction site on our screens. In just nine days, a hospital had been built, with a thousand beds and thirty intensive therapies. While carpenters and masons were still completing the hospital's rear area, the first patients worldwide, for a serious and mysterious disease, were being admitted to the front area in droves. At this point—as reported by Peter Hessler in *The New Yorker*—many of the construction workers fled without bothering to wait for their salaries, which, given the emergency, were up to ten times higher than normal. Their terrified faces offered up a small taste of how life on the planet would suddenly and radically change, thanks to a tiny, invisible virus.

As we write this book, more than two years have passed since January 2020 when some Chinese researchers from the University of Fudan published online the genome sequence for a virus that was, as yet, unknown to the human immune system. They did so courageously, at a personal risk of reprimand and punishment by government officials.

The virus was isolated from the swab carried out on a Chinese patient who worked in a Wuhan food market. One of the first known outbreaks of the infection started there, at the end of 2019. By now the virus has been named SARS-CoV-2. Its genome, protected by a membrane and an envelope made of proteins and lipids, comprises a single strand of RNA of about thirty thousand nucleotides. The new coronavirus has been identified as the causative agent of COVID-19 disease (in full, coronavirus disease 2019), and of the pandemic that is still ongoing.

Much has been written about this disturbing news and how it was held back by officials afraid of displeasing the authorities. Had this information circulated earlier, the epidemic might have been stopped in its early days. Much has also been said about the Trump administration's inauspicious decision to withdraw the United States, just before the outbreak, from the long-standing collaboration with China on surveillance of epidemiological and pandemic risks: a hostile move that has possibly facilitated the frenzied race of SARS-CoV-2 around the planet.

It is plausible to hold that things would have gone differently in an ideal world. But it is also possible that a fast, contagious, and deadly virus like SARS-CoV-2 would still have raged, even in the more just, collaborative, and harmonious world we all would like to live in.

In any event, crying over spilled milk is not very useful. From the desolate landscape where the COVID-19 pandemic has now placed us, we can instead try to understand why this virus should not have taken humanity by surprise. We will look at how we tried to protect ourselves and what lessons we can take from it all, for the future.

## The COVID-19 numbers

As we finish writing this book, as of late February 2022, the virus has infected more than 433 million people and killed over 5.9 million around the world, according to data collected by the Coronavirus Resource Center at the Johns Hopkins University. More than 18.9 million cases and over 161,000 deaths have been registered in the United Kingdom alone. These are cases and deaths officially attributed to COVID-19, based on tests, diagnoses, and tracing. Many epidemiologists estimate that real numbers are at least 10 times higher for cases and 2 or 4 times higher for deaths, while recognizing the very high uncertainty of such numbers.

Comparisons with the "Spanish" flu pandemic of 1918 have been frequent, given the ferocity of both viruses. According to some estimates, the "Spanish" flu killed more than forty million people in less than a year, out of a world population that was a quarter of todays. These numbers are a shocking reminder of how vulnerable, defenseless, and deprived of liberty a world can become when there are no effective vaccines against a lethal virus.

COVID-19 has also reduced the life expectancy of the world population, as the "Spanish" flu did then. According to a study conducted by Oxford researchers, scientists have estimated that in England and Wales the pandemic has cut life expectancy by roughly a year, reversing gains made since 2010. A similar decline had not been recorded since the two World Wars.

## The SARS-CoV-2 virus

The SARS-CoV-2 virus is spread via droplets emitted by infected people, not only by coughing and sneezing, but also by simply speaking and breathing. The droplets emitted can disperse in the air and persist for a few hours, especially indoors, before falling to the ground or evaporating. Because of this, the virus is said to be airborne, and defined as such by the American Centers for Disease Control and Prevention and the WHO.

With the onset of infection, the virus penetrates into the host's cells thanks to the S glycoprotein. The letter S stands for the now notorious spikes depicted in many illustrations of the virus. The viral particles attach themselves to a receptor on the surface of a human cell, thanks to the Spike protein's S1 subunit. The binding initiates fusion of the viral and cellular membranes, enabling the viral genetic material to enter into the cytoplasm. The human cell receptor that acts as a gateway for the virus is specific to the angiotensin-converting enzyme 2 molecule, and is present on the cell membranes of the respiratory tract, throat, and lungs, where the virus penetrates from the outside.

Once inside the cell, the viral RNA takes control of the cell's synthetic machinery, which starts producing all the pieces necessary to assemble new viral particles at maximum speed. Once the new viruses are assembled, they are quickly transported out of the cell. Here, the freshly "baked" viruses can infect other cells of the same individual or of other people when released in the droplets dispersed by an infected person's breath: a single infected cell can produce millions of viral particles. Observed under an electron microscope, SARS-CoV-2 shows a "crown" on its surface that is typical of other coronaviruses (Figure 1).

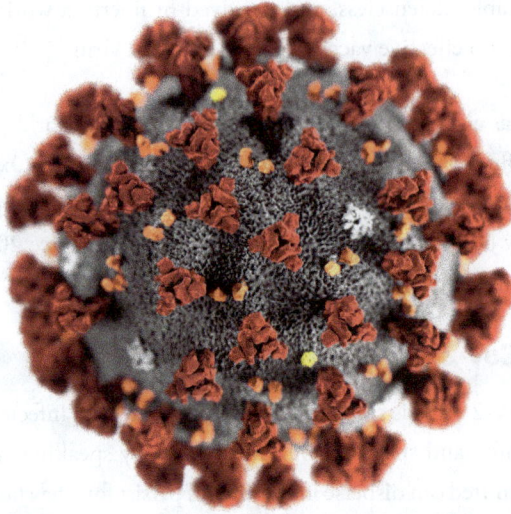

Figure 1: The SARS-CoV-2 coronavirus in the illustration created by the US CDC.

## COVID-19 syndrome and its complicated consequences

So far what we have learned about a disease that was unknown until two years ago, is that the infection leaves roughly 50 percent of people without any symptoms, following an incubation that can last from two to fourteen days.

In people with symptoms, the first phase of the disease lasts about a week, with the virus actively reproducing itself, especially in the respiratory tract or intestines. The most common symptoms for this viral phase are bone pain, fatigue, dry cough, fever, temporary loss of smell and taste, and sometimes intestinal problems.

These symptoms then subside, with most people recovering, but some relapse after an initial improvement. For these patients the main problem is no longer the virus, but a serious state of inflammation, due to a confused, faulty, and persistent reaction by the immune system. This type of reaction can last several weeks or even months and can go so far as cause severe or

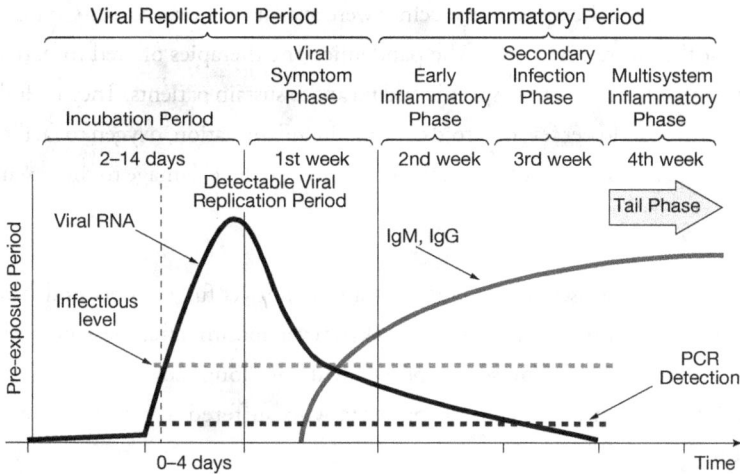

Figure 2: The graph summarizes the viral and inflammatory phases of COVID-19 disease (Daniel Griffin, Columbia University).

even fatal damage to the lungs, circulatory system, heart, or other affected organs (Figure 2).

It is especially in this second phase that most seriously ill patients are hospitalized. Many eventually recover, but about 0.15–0.28 percent of the total cases recorded globally have been fatal. Most deaths registered are among the elderly and people with concomitant diseases, whereas the least susceptible are the younger members of the population. Still, without vaccines no one is protected from COVID-19, and even young adults have been dying of the disease at an impressive rate, unheard of in the last 100 years. According to data published in the *Journal of the American Medical Association*, between March and July 2020, almost 12,000 more people have died in the United States within the 25–44-year age group than in previous years. Among the survivors, many risk remaining seriously injured for life. As Lawrence Wright reported in *The New Yorker*, Chris Rogan was a young and healthy 29-year-old American employee. In August 2020 he was discharged from a New York hospital with one leg amputated above the knee, the result of damage caused by the disease, along with many other per-manent injuries, following a five-month COVID-19 hospitalization.

Neither specific therapies nor vaccines were available against SARS-CoV-2 for almost the entire first year of the pandemic. The therapies offered to patients during that time were non-specific and meant to sustain patients. They included medications to lower fever, steroids to contain inflammation, oxygen to facilitate breathing, and anticoagulants such as heparin to reduce damage to the vascular system.

The long-term consequences of the disease are not yet fully understood. Some people show symptoms that last for weeks or even months after the acute phase. The syndrome, called by some "post-Covid" or "long Covid," is still being studied and can also occur in patients who suffered a very mild initial disease.

## Contagion prevention and lockdown

In the spring of 2020, the world air traffic fell silent as wild boars strolled through the deserted streets of big cities, like Barcelona, disturbed only by the incessant sound of ambulance sirens. During those months, when the first lockdowns were imposed in thousands of places in the world, many frightened employees tried to familiarize themselves with remote working amid the tantrums of bewildered children at home from school. Many other workers were suddenly left without a job and salary, while an even more dramatic collapse of society was prevented thanks to the so-called "essential" workers like supermarket clerks, truck drivers, doctors, and nurses.

The virus and its tragic consequences have upended our habits, changing our lives as we knew them while depriving us of our freedom. The lockdowns have been extreme and desperate measures, taken when no vaccines or specific and effective therapies were available against this very contagious, airborne microbe.

We initially tried to protect ourselves by physical distancing from other people, by adopting masks, and other protective measures, like improved hygiene, diagnostic tests, and contact tracing of people who tested positive. However, when these measures proved insufficient and as hospital wards were being filled beyond capacity with too many sick people, more stringent lockdowns became inevitable in many countries. Unsustainable in the long run, these extreme attempts to avoid an even greater numbers of deaths, due to the collapse of health systems, have led to overwhelming economic and social costs, which are no doubt incalculable in full.

The sadness, the desolation, and the regret for the many lives cut short by the pandemic have been accompanied by shock and dread at the impressive speed with which the virus has spread worldwide, like a tsunami. The waves that have followed, from one continent to the next, have upset the existence of over 7.5 billion people, pretty much the entire world population. Until July 2021, with the exception of one small island nation in the South Pacific where the virus appears not to have arrived—Tuvalu—there was no country that had not registered cases (the regimes of North Korea and Turkmenistan, who have not reported infections, presumably owe the apparent immunity of their respective populations more to censorship than to effective containment measures).

## Democracy and the Swiss cheese model

To reduce the risk of coming down with COVID-19, public health officials told us to maintain a distance of at least six feet from other people, but how do you do this in a two-bedroom apartment crammed with five to six people? We were told to cover our nose and mouth with masks, but even the most effective masks do not offer 100 percent protection. Also, at the beginning of the pandemic, masks were not even available for healthcare professionals, and the confused messages we received at the time contributed to the refusal of a necessary but annoying measure by a part of the population. We were asked to stay at home and isolate ourselves in case of fever and other symptoms, but how many people did not understand the rule, or simply could not afford to follow it? We were shown how to disinfect our hands properly, but how many places are there in the world where neither disinfectants nor running water and soap are available? We were advised against touching our faces, but how many times does one do this inadvertently? We were told to avoid crowds, but do we really have a choice when a bus or a train is packed, when we need to go home from a job that supports our family, and we do not own a car? They urged us to spend more time outdoors than indoors, but how do you do that in countries where winter is long and bitterly cold?

Health systems were required to test as many people as possible, to trace the contacts of those who tested positive for the virus, to isolate noncritical patients, and to quarantine suspects. However, molecular tests are only accurate and reliable up to a point (the best ones up to 70 percent), and even highly efficient Germany, with its own contact tracing, has struggled to keep up with the speed of contagion of SARS-CoV-2.

Doctors and nurses were asked to protect themselves, avoiding the spread of the virus to patients, but in many countries supplies of emergency equipment had not been replenished. Several public authorities had even forgotten to update their pandemic plans, resting on the laurels of earlier antibiotic and vaccine successes. The threat of infectious diseases has long been considered a nuisance rather than an imminent and lethal risk.

Even more recently, building superintendents, homeowners, and employers have been urged to better ventilate homes and offices, but how many air exchange systems are state of the art and properly maintained? Some governments have done their best to support workers and businesses that have suffered the most from the pandemic with economic aid. But when the need is immense and the "blanket" is really too short, how do you prevent public anger, rebellion, protest gatherings, and subsequent new clusters of infections?

Only a rich, powerful, efficient, and undemocratic country like China seems to have at least partially managed to push the virus out of its borders. However, the goal has been accomplished thanks to authoritarian and draconian measures that had not been seen since the time of Mao Zedong. During the initial lockdown in early 2020, in the city of Wuhan alone the Chinese government sent 1,800 teams of epidemiologists, each consisting of at least five people, to contact-trace infected citizens. As Peter Hessler wrote in *The New Yorker*, in 2020, "the Chinese lockdown was more intense than almost anywhere else in the world. Neighborhood committees, the most grassroots level of Communist

Party organization, enforced the rules, and in many places they limited house-holds to sending one individual outside every two or three days to buy necessities. If a family were suspected of exposure to the virus, it was not unheard-of for their door to be sealed shut while tests and contact tracing were being conducted."

In parts of the world, where personal freedom is a non-negotiable value, the weaknesses of each layer of defense have shown their limits. In this regard, the Australian virologist Ian Mackay has proposed the so-called "Swiss cheese model of pandemic defense." Figure 3 shows, with a metaphor, that no single intervention is perfect for preventing the spread of the coronavirus and that each one has holes. The idea is that, when the holes align, the risk of infection increases. But combining several layers, with holes left imperfectly aligned, creates more obstacles to the passage of viral particles. It should be noted that only scientifically validated measures of risk limitation were taken into account in the Swiss cheese model. This does not include, for instance, sanitation based on the indiscriminate use of chemical products, so widely deployed in this pandemic.

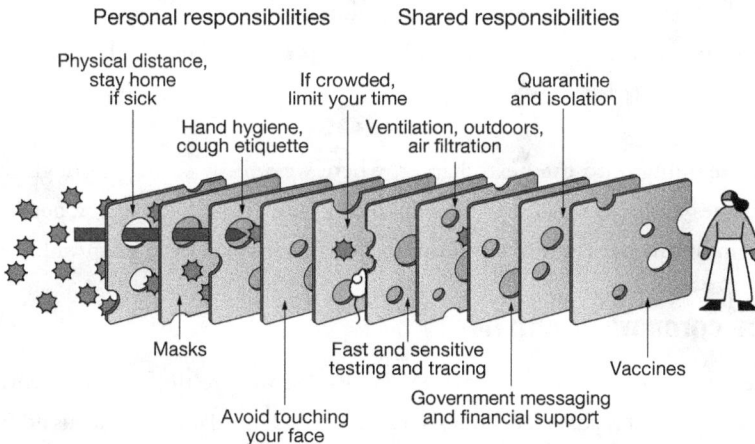

Figure 3: According to the "Swiss cheese model of pandemic defense," no single preventive intervention alone is able to block the spread of the coronavirus and each layer has holes. However, several combined layers of intervention, including vaccination, can create serious obstacles to viral contagion (adapted after Ian Mackay).

The last slice in the Swiss cheese model is the one that involves vaccines. Its position, at the end of the drawing, tells us that there is no way to end the COVID-19 pandemic without immunizing a large part of the world population with specific and effective preparations.

In the next chapter we will be looking at which types of SARS-CoV-2 vaccines have been developed in record time, thanks to the preparation and efforts of the entire global community of vaccinologists and the massive resources made available by many national governments. These are vaccines that, despite the extreme emergency, have taken 10–11 months from vaccine design to administration of the first doses. It took a further year before the vaccines could be administered into the arms of a substantial part of the population, at least in wealthier Western countries. As we go to press, the vaccination campaign is still ongoing and it will very likely continue, possibly with additional COVID-19 vaccines, in the years to come.

Until a widespread protection can finally marginalize the virus, we have no choice but to swallow these indigestible morsels of "Swiss cheese." Aware of all the holes in its layers, we must patiently adapt to the protective measures. They are the best we have to avoid too many additional losses of human lives to the virus, before the situation can gradually improve and ultimately, thanks to vaccines, be settled.

Before going onto the next chapter, where we will be sorting through the vaccines available for this pandemic, let us take a brief look at the possible origin of this virus, and the conditions that may have favored its spread.

## The coronavirus and us

We owe both the term "coronavirus" and its discovery to the Scottish scientist June Dalziel Almeida. Her debut seems far-fetched: the daughter of a bus driver, at sixteen she was forced to abandon her studies for lack of money. She then became a laboratory technician and studied biology on her own. After moving to Canada, she developed a pioneering immunoelectron-microscopy technique which, thanks to the use of antibodies, allowed for a higher resolution of virus imaging. Returning to England in 1964 she identified, in a laboratory at

St. Thomas's Hospital, in London, a new type of virus that was showing a halo similar to a crown. It was a virus that had been isolated from the nasal discharge of people sick with a cold. The discovery was initially snubbed by colleagues, who thought that the photographs had not been taken properly and that the fuzzy images were of out-of-focus flu viruses.

Hundreds of different types of coronaviruses have since been discovered and isolated. They are all RNA viruses and many affect animals that humans can come into contact with, such as camels, cats, chickens, and bats. Among the types that are capable of infecting our species, four are responsible for about a third of common colds, while three, identified more recently, can prove fatal: SARS-CoV, MERS-CoV, and SARS-CoV-2.

Prior to the close of 2002, when the SARS-CoV virus suddenly emerged in China, coronaviruses were not very exciting for virologists, and those dealing with them were quite on the fringe of the discipline. This all changed suddenly with the first SARS epidemic. SARS was a serious respiratory disease whose causative agent, the SARS-CoV virus, rapidly infected more than 8,000 people in 26 countries, killing nearly 800. After alarming the public health authorities worldwide, the virus seemingly disappeared in 2004.

The virus was isolated thanks to the work of Carlo Urbani (Figure 4), an Italian doctor and WHO consultant for the control of parasitic diseases in the Western Pacific. Urbani quickly recognized that the atypical pneumonia of a patient hospitalized in Hanoi, Vietnam, was in fact a new disease and he raised the alarm with the WHO. He also managed to convince the Vietnamese authorities to impose a lockdown and quarantines that were crucial for the containment of the virus. Tragically, Urbani himself became infected, and the disease killed him in 2003.

Less than a decade later, in 2012, the MERS-CoV virus, responsible for MERS, was isolated and identified in Jeddah, Saudi Arabia. The virus had been found in dromedaries in several countries of the Middle East, Africa, and South Asia. Based on the analysis of several viral genomes, it is likely that the MERS-CoV virus was initially confined to bats before jumping to camels, at some unknown

Figure 4: Dr. Carlo Urbani (Archive of the Italian Carlo Urbani Association).

point in time in the distant past. Since 2012, it has continued to cause sporadic local outbreaks, infecting small numbers of people. It is highly probable that contagion occurs through contact with animals such as dromedaries, camels, or bats, which make up the virus' animal reservoir. A total of twenty-seven countries have reported cases, resulting in 888 known deaths from the infection and its complications. The small numbers confirm that transmission of MERS-CoV between humans is possible, but rare.

The experience gained with the SARS and MERS epidemics has better prepared the countries affected, such as Vietnam, Taiwan, and Hong Kong, to cope with COVID-19. Compared to the rest of the world, the governments and public health authorities there were more aware of the danger posed by the virus and their decisions were timely and effective in limiting contagion and deaths.

## Spillover

It is possible that SARS-CoV-2 entered the first human body from a wild or farm-raised animal, as its predecessors probably did. The jump could have been direct or through an intermediate host: in other words, the virus could have spread from one of the many species of horseshoe bats that abound in nature in southern China, either through direct contact with a human, or after first spending some time in another species.

Recent studies have established that the SARS-CoV-2 genome resembles that of RATG13, a coronavirus first isolated in 2013 in a horseshoe bat in the southern Chinese province of Yunnan. The RATG13 genome is approximately 96 percent identical to the SARS-CoV-2 genome. This percentage of identity might appear very high, but humans and chimpanzees also have approximately 96 percent identical genomes, with significant differences between the two species. This suggests that, like primates, SARS-CoV-2 and RATG13 may have shared a common ancestor, perhaps a "great-great-grandfather" dating back many years, with the "great-grandfather," "grandfather," and the "father" of SARS-CoV-2 still unidentified.

As David Quammen pointed out in his article in *The New Yorker* in 2020, scientists have yet to discover exactly which fateful encounter brought this coronavirus into the human species, and it is possible that they never will. Still, viruses do not actively chase us: they do not fly, climb, or move on their own, but migrate inside the bodies of other animals which, if left alone, show little interest for humans.

A spillover occurs when a pathogen enters into a human being for the first time, jumping from an animal species that is a reservoir for the virus, in this way broadening its range of prey.

Spillovers have taken place since the dawn of humanity, but nowadays we are creating more opportunities for spillovers whenever we encroach upon other species' habitats. With bats, this can happen when we dig into their guano for fertilizers, or catch and kill them, or transport them live to markets, or interact with them in some other disruptive way.

About 70 percent of emerging infectious diseases and every pandemic over the last twenty years have originated from the complex interactions between individuals of our species and wild or farmed animals. For instance, the Nipah virus outbreak that hit Malaysia in 1998 was favored by the rapidly growing number of pig farms and pork processing plants. The hastily built facilities were located near tropical forests, colonized by large populations of wild bats. SARS and Ebola, which emerged in China and Africa respectively, have also been connected to large colonies of wild bats: the poor and hungry communities living nearby had had to resort to hunting them to feed themselves.

However, microorganisms have been jumping from species to species for millennia, long before we were capable of large-scale environmental and eco-logical disruption. Perhaps the deadliest virus that has passed from bats to humans is rabies, a disease we have known of since at least the time of Democritus, in the fifth century BCE (We already encountered rabies in Chapter 2). Without immediate vaccination after exposure, the mortality rate for rabies is almost 100 percent, with the disease still killing tens of thousands of people every year worldwide.

In 1911, Antonio Carini, an Italian doctor and microbiologist working in Brazil, reported the discovery of the rabies virus in bats. Carini also noted that the virus did not appear to make these winged mammals sick, suggesting that the relationship between the bats and the virus had been long lasting. Possibly the two species had adapted to each other in a mutually favorable way: a safe habitat for the virus and no viral symptoms for the bat. Carini's discovery was, in fact, a strong indication that originally the rabies virus might have passed from bats into the body of a dog or other carnivore, and that the saliva of one of their descendants had then dripped into the bite wound of some unfortunate human.

We call the animal diseases that are transmissible to human beings *zoonoses*. Although zoonoses have always existed, some, including the plague, the flu, AIDS, and now COVID-19, have caused some of the deadliest epidemics in history, and their recent increase is a cause for worry. In order to feed and sustain itself, the enormous and ever-increasing human population needs to deforest progressively larger regions of the planet, expanding cultivated areas, intensifying livestock farms, and constructing new homes and factories. Animals, which previously lived in these now degraded environments, either become extinct, or migrate, or they start to live in contact with us. But it is our behavior, alongside powerful technologies, that ensure that this all takes place with a rhythm and impact that were unimaginable even fifty years ago.

## Escaped from a lab?

Another hypothesis, which is less likely, but not completely ruled out, is that SARS-CoV-2 may have escaped from a laboratory because of a human error. The virus might have been studied by researchers for its potential for human contagion, after being isolated in wild animals or in a farmed species.

This rumor has been circulating since the beginning of the pandemic. During the Trump administration, the American secret services even declared they had supporting evidence for this theory. However, no data have so far been made available to the scientific community.

Investigations carried out by experts appointed by the WHO failed to disclose any overwhelming evidence in favor of either hypothesis: whether the more likely natural origin of the virus, or the more unlikely accidental leakage from a laboratory.

The scientific community agrees on the need to continue the enquiry, for a better understanding of what happened and to protect ourselves in the future. To be worthwhile, the investigations should receive the frank and sincere cooperation of experts from all nations involved and should be sheltered from the political rivalries and conflicts between nations.

This will also require a lot of patience. Investigations into the origins of an epidemic take years and are often unsuccessful. For instance, it took fourteen years to establish that the SARS epidemic, in 2003, was started by a coronavirus that jumped from bats to humans, most likely through an intermediate passage in civets.

## Our planet's health is ours too

If it is so dangerous for the individual and for society to encroach upon the existence and lifestyles of other animals, why do we insist on disturbing a myriad of delicate natural balances? The ultimate cause of these behaviors is the rapid growth of the planet's population, which has generated an inexhaustible and ever-increasing demand for energy and food. For this reason, humanity, especially in the latest generations, is consuming natural resources at a pace that is unsustainable for the planet. The consequence is that plants, animals, and entire ravaged ecosystems are no longer able to regenerate on their own, while the climate undergoes rapid transformation.

The impact of these actions is evident above all in the extinction of millions of species at risk and in the irreversible transformation of environments, such as glaciers that we had previously considered immutable. Less visible, but no less devastating, are the consequences of these upheavals on microorganisms and on the dangerous infections they can potentially spread. A disturbed ecosystem offers the most dangerous microorganism's opportunities: something that, for the most part, has been ignored. This lack of consideration is possibly due to the invisibility of this danger for an inexperienced and untrained eye.

If, on top of all else, we add the pressure exerted by climate change on the poorest populations, with the resulting conflicts and migrations, the consequence is a highly disturbed natural environment, with the entire human species becoming more vulnerable and exposed to new infections. In the past, an outbreak that emerged from an abused Asian or African forest tended to remain confined to the region and its local population. This is no longer the case today. With millions and millions of people travelling fast the world over, every day, myriads of microscopic and unwanted guests are being carried around. Most of the time, they are impossible to detect and are blocked, for instance, by measures of containment, or by social distancing, vaccines or drugs when available, or by any other available means.

It is also a fact that pockets of extreme poverty and neglected diseases are frequent today, even in very rich countries. In 2011, Professor Peter Hotez had just moved from the East Coast to Houston, Texas, to head the National School of Tropical Medicine at the Baylor College. There he discovered that many diseases, normally present in developing countries only, were in fact widespread in the poorest neighborhoods of his new city. The illusion that the populations of the wealthiest nations are living in gated communities, sheltered from the diseases and contagions of the poor world, is over.

## Our intimacy with animals

Our ancestors have cohabited with animals for most of the history of our species. Even if we have mostly forgotten it, this close cohabitation has been going on even in rich countries until quite recently, at least until our grand-parents' generation, sharing all kinds of microbes together with those animals that have lived alongside us.

As the anthropologist Jared Diamond pointed out, the measles virus is, for instance, a close relative of the rinderpest virus which causes cattle plague, a serious disease affecting cattle but not humans (while measles itself is not transmitted to cattle). It is possible though that, at some time in the past, a strain of rinderpest virus mutated into a measles virus, an adaptation that helped it thrive in our bodies. Considering that many farmers still live next to their livestock, along with animal feces, urine, blood, and saliva, this should not come as a surprise. "Our intimacy with cattle has been going on for the nine thousand years since we domesticated them—ample time for the rinderpest virus to discover us nearby," writes Diamond.

It is not that long ago that we moved cows and calves from inside our homes, relegating them to farms away from our sight. However, we do not shy away from the company of other animals (and their germs). Monkeypox, for instance, appeared suddenly in the United States in 2003, with an outbreak of seventy-one cases, as the British virologist Dorothy Crawford said in a lecture in 2010. The virus has long been known to infect rodents in Africa (and despite the name, it never had anything to do with monkeys). Monkeypox had been imported into the United States from Ghana, inside the body of a giant Gambian rat that someone, curiously, had acquired as a domestic animal, alongside cats and dogs. The Gambian giant rat was initially housed in a pet shop near some prairie dogs that became infected. Once sold, the prairie dogs began to get sick, along with their owners too.

When we ask the poorest inhabitants of the planet to catch animals that some of us want to eat, or keep in our homes as pets, we probably do not realize how we are, in fact, offering the viruses that accompany these animals with unexpected opportunities for travelling and finding new hosts.

## Cruise ships or slums?

The outbreaks that occurred on several cruise ships crammed with passengers at the beginning of the COVID-19 pandemic caused a media sensation. No country was giving permission to dock those floating buildings, which suddenly transformed into sort of quarantine hospitals. Crews and passengers were forced into confinement on the ships for long periods of isolation.

"You don't have to be very poor to live with the population density of a slum," said Dorothy Crawford, commenting on the fact that many wealthy people in Western countries love to spend their holidays in such crowded environments. Small cabins with poor ventilation, with guests lodged in very crammed spaces, are also the beloved dwelling for several viruses. Noroviruses, for instance, often ruin the holidays of cruise passengers, leaving them with legendary diarrheas and vomiting that are, in turn, responsible for spreading the infection to other guests, possibly through poorly insulated air and water pipes.

Conditions such as these are not unlike the ones we have created in many crowded environments of our modern, affluent life. In open-plan offices, scores of employees share the same air for many hours a day, air that is often made stale by sealed windows and ventilation systems that do not always guarantee an adequate air exchange. It is as if Western society, spoiled by just a few decades of freedom from infectious diseases, had forgotten simple hygiene and prevention measures that have been essential to the survival of our species for centuries.

## Vaccines are an ecological and "green" solution for the elimination of infectious diseases

At least one billion species of microorganisms, mostly bacteria, inhabit our planet. At any given moment there are five million trillion trillion ($5 \times 10^{30}$) living bacteria with $10^{31}$ bacteriophages, or bacterial viruses, constantly attacking them, killing roughly 40 percent of them every day. Overall, microbes are the driving force of evolution. To match their daily mutation rate, we would need no less than ten thousand years of laboratory experiments.

Microorganisms were the first living beings to appear on Earth, about 3.8 billion years ago. The first eukaryotic cell, from which all animals and plants evolved, humans included, was a microbe. We need microorganisms to produce many basic foods, such as bread, wine, and beer; to decompose the abundant sewage that we pour into water treatment plants; and to produce essential drugs, such as antibiotics.

At least one hundred trillion microbes live inside the intestines of every human being, contributing to essential functions like metabolism and immunity. A similar symbiosis is present in other animals. The vast majority of microbes are therefore useful to humans and the environment. Only a tiny minority, around 1,400 kinds of bacteria and viruses, out of an estimated total of millions of trillions, are responsible for infectious diseases that are dangerous for our health.

Natural selection has favored those living beings able to evolve tricks at any point of their life cycle for avoiding pathogenic microbes. Despite this, germs are almost always ahead of their hosts: they have a higher degree of genetic variation; they are faster at changing shapes and functions, and they have a huge selection of live matter from which to test each new variant for its infectious and reproductive fitness.

Every day we clean houses and offices against these pathogens, using chemicals that kill almost any microbe. Detergents and disinfectants are not very specific. They destroy everything they encounter indiscriminately, a bit like the napalm used in the Vietnam war. Also, at the slightest sign of infection, we are prescribed broad-spectrum antibiotics. These drugs not only eliminate the bacteria that presumably caused our disease, but all other bacteria in the body, many of which are in fact beneficial.

This coarse attempt to eliminate the approximately 1,400 species of pathogens is jeopardizing the survival of the remaining millions of trillion bacteria and viruses. The ecosystems that depend on them for the support of all forms of life on the planet, humans included, are at risk.

Add to this the hundreds of thousand tons of antibiotics administered yearly to chickens, cows, and other farm animals to accelerate their growth. These drugs are then discharged into the environment through sewage, where their persisting effects continue on the microorganisms normally found in waters and soils. As a consequence, microbial diversity has been greatly reduced while we witness the appearance of superbugs that are resistant to any treatment, as we will see in Chapter 9.

Vaccines, on the other hand, are modelled on the natural properties of our immune system. With over millions of years of evolutionary experience behind it, our system has learned to keep the microbes it encounters at bay in a more targeted way, without affecting beneficial microorganisms. Vaccines are very specific too, intercepting only those pathogens that they were designed to fight, without affecting other microorganisms. With vaccines, the microbial diversity of both the organism and the environment are kept safe. This makes vaccines an ecological and "green" solution for the eradication of infectious diseases, COVID-19 included.

# 8

# How to Vaccinate an Entire Planet Against COVID-19

Most of the vaccines you have read about in this book took years, if not decades of research to be developed and tested, before being approved and administered to the general public. In 2020, scientists in many countries were able to develop safe and effective vaccines against the new, very contagious, and lethal coronavirus, and to test the vaccines in record time.

The SARS-CoV-2 genome sequence was published online in January 2020. Less than eleven months later, on December 8, a ninety-one-year-old English woman received the first dose of an effective and safe vaccine against COVID-19 that had just been approved for emergency use by the strict regulatory authorities of a Western country.

The "Covid vaccine cavalry," as Anthony Fauci called it, had arrived to save us, well before expert forecasts, which had estimated at least 2–3 years. By February 2022, 17 vaccines had been approved globally for early or limited use, and 12 for full use, as vaccination campaigns were being rolled out worldwide, although predominantly in wealthier countries. At the same time, 115 vaccines were still being tested in human volunteers; 49 of which had reached the third and largest clinical trial phase, with at least 75 other vaccines under study in laboratory animals. These impressive numbers testify to the efforts and creativity deployed in every corner of the world, from the United States, to China, to Europe, to Russia, as well as many other countries, to stop the scourge that has hit us. A complete list of all COVID-19 vaccines would take up several pages of this book and be obsolete even before going to print. For an overview and a complete Covid vaccine list, updated in real time, you can check the "Coronavirus Vaccine Tracker" on the *New York Times* website

(*The New York Times* was awarded the Pulitzer Prize 2021 for its excellent coverage of the pandemic).

## Vaccines against COVID-19

What are the main features of COVID-19 vaccines? And how was it possible to produce them so quickly?

First of all, vaccinologists were cheered to discover that making vaccines against SARS-CoV-2 is a relatively easy process, at least in comparison to other kind of viruses, like HIV or HCV.

The vaccines developed and tested against COVID-19, with varying degrees of success, belong to four categories. The fastest to develop, and the most innovative, are those made with mRNA, where the only initial requirement is knowing the viral genetic sequence. The second fastest are vaccines based on adenoviral vectors: in addition to knowing the SARS-CoV-2 sequence, their development requires an adenovirus that can be modified in the lab and used as a vehicle to transport the vaccine into the cells. More time is needed to develop vaccines that are based on recombinant proteins or inactivated viruses, as they require more laboratory steps to produce the antigen, or to grow the virus in cell culture or in eggs.

A description of the characteristics for the main COVID-19 vaccines is set out below in greater detail, along with their efficacy data, where known.

### mRNA vaccines

These are entirely synthetic vaccines, built from a completely new concept. The idea of using mRNA in vaccines, first conceived in 1992, became real in 2005, thanks to the pioneering perseverance of Kati Karikò and Drew Weissman, when both worked at the University of Pennsylvania. Their method, at least in the early stages of the process, was a great time saver.

The researchers began by inserting a synthetic gene, whose original counterpart was present in the viral sequence, into a plasmid. The synthetic gene contains a slightly modified version of the instructions to build the so-called S glyco-protein that gives shape to the now notorious spikes. The virus uses the spikes to attach itself to a receptor on the surface of a human cell, allowing the viral genetic material to enter the cytoplasm.

The genetic instructions to build the S glycoprotein have been slightly altered in the vaccine, in order to obtain a more stable version of the protein: one that is better recognized by the immune system prior to fusion with the cell membrane. We already mentioned this strategy for vaccine development in Chapter 4, regarding the RSV vaccine. Jason McLellan and Barney Graham, at the NIH in Bethesda, Maryland, were involved in the design of this part of the mRNA vaccine, in collaboration with the manufacturers. The two research-ers designed the instructions for the modified S glycoprotein, for insertion into the vaccine, just one day after receiving the genetic sequence of the SARS-CoV-2 virus and in this, they were aided by their previous work on a prototype vaccine against MERS-CoV. Once the synthetic gene had been inserted in the plasmid, in less than a week they were able to obtain an mRNA molecule, built with modified nucleotides to be better tolerated by the human immune system.

The mRNA molecule in the vaccine is wrapped by a nanoscale lipid envelope, which serves both to protect the mRNA from deterioration, and to facilitate transport into the cells.

Inside the cell, the vaccine mRNA instructs our machinery for protein synthesis to produce the stabilized S glycoprotein, which is immediately exposed, both whole and in pieces, on the surface of the cell itself.

At this point, the cells of the immune system take note of this unknown protein and its parts, and begin to memorize its features and produce antibodies, in readiness for an infection.

These kinds of vaccines have also been called "digital" because their design, development, and production only requires information that travels online and in synthetic, non-biological materials. From this point of view, the pro-duction of so-called "digital" vaccines is radically different from the methods

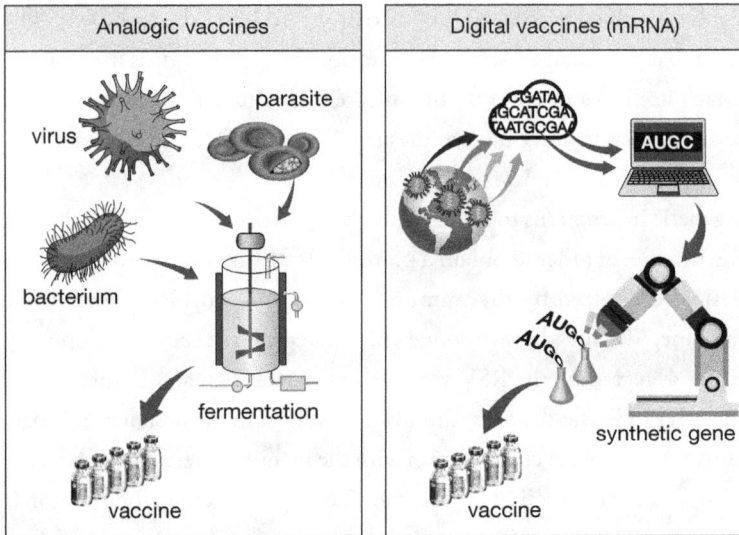

Figure 1: On the left, the scheme of conception and development of a traditional vaccine, so called "analog" compared to the entirely "digital" system with which an mRNA vaccine can be produced (Rino Rappuoli).

used to develop more traditional, "analog" vaccines, which instead require the germ to be manipulated and grown in a fermenter (Figure 1). The other big difference when compared to traditional vaccines, is that in this case the injected material is not a finished vaccine, but the instructions (in the form of mRNA molecules) for our cells to make such a vaccine on their own.

Two mRNA vaccines have been developed and produced by two biotech companies, Moderna in the United States and BioNTech in Germany, in collaboration with the US pharmaceutical industry Pfizer. Both vaccines have been tested in tens of thousands of volunteers, passing safety checks and demonstrating an efficacy of over 90 percent after two doses. Initially approved for emergency or conditional use in the United States, in Europe, and in many other countries, mRNA vaccines have now immunized the citizens of hundreds of countries around the world.

Believing and investing in this type of vaccine was a bold bet, because no mRNA vaccine had ever been approved before COVID-19. So far, the bet has proven a winner, even if, at least in the early stages of the vaccination campaign, its

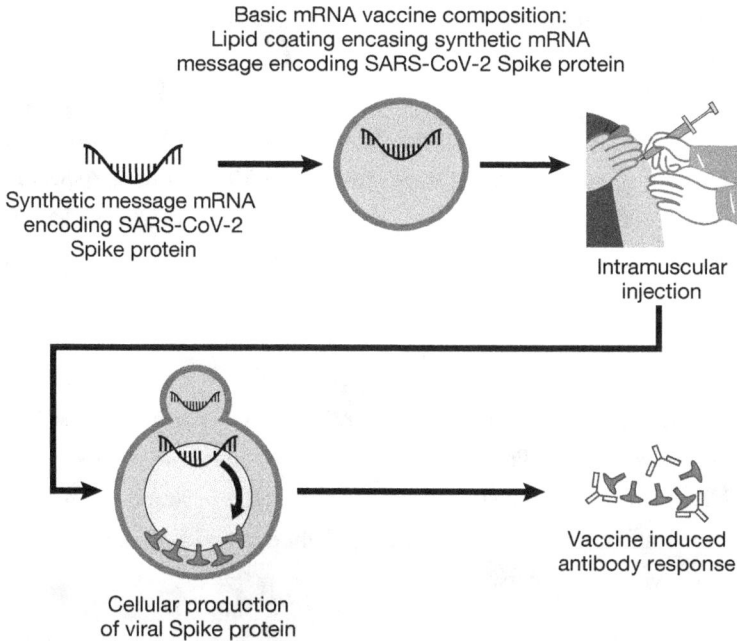

Figure 2: How a COVID-19 mRNA vaccine works (adapted from the *American Society for Microbiology*).

manufacture required adjustments as well as the construction of new bioreactors to keep up with demand, of the order of billions of doses.

In the future, some improvements would be welcome to ensure a simpler and more extensive use of this new technology. One shortcoming of these type of vaccines is that they need to be stored at very low temperatures, respectively at −70°C for Pfizer-BioNTech and at −20°C for Moderna, because of the instability of the lipid envelope at higher temperatures (Figure 2). The German mRNA vaccine CureVac seemed to meet some of these requirements, although it is efficacy in trials against SARS-CoV-2 and variants, at around 48 percent, was considered too low to proceed with the regulatory approval process.

## Vaccines with adenoviral vectors

Here, a synthetic double-stranded DNA gene with instructions for manufacturing the SARS-CoV-2 S glycoprotein, is inserted inside an adenovirus that is unable to reproduce in humans and rendered further harmless through appropriate genetic modifications.

Adenoviruses are common viruses that cause colds or flu-like illnesses in various animal species. Those used in COVID-19 vaccines are specific to species other than human beings. When the modified lab-grown adenovirus is injected into the body, it can penetrate human cells but cannot reproduce itself.

Once the vector enters a cell, the synthetic gene is first transcribed into an mRNA molecule, and then translated inside the cytoplasm into the SARS-CoV-2 S-glycoprotein. From then on, the sequence of events is the same as for mRNA vaccines: the antigen is exposed on the cell surface and spotted by the immune system, which starts producing antibodies and storing memory cells for future viral encounters.

Perhaps the most famous vaccine developed with this method is the one designed at the Jenner Institute, University of Oxford, in the United Kingdom, in collaboration with the pharmaceutical company AstraZeneca and tested in tens of thousands of volunteers. Effective at approximately 62 percent, it has been initially approved for emergency or conditional use in the United Kingdom, in the European Union, and elsewhere. The advantages of this vaccine are its very low price, only a few pounds per dose, and the greater ease of storage and transport, since the preparation remains stable in a normal refrigerator for roughly six months. These are crucial factors for use of the vaccine in poor countries.

Then there is the vaccine from the Chinese company CanSino, whose efficacy is about 65 percent, the Sputnik V vaccine from the Russian Gamaleya Institute, possibly protective in more than 90 percent of cases (not all data are known and published), and the Janssen vaccine, manufactured by Johnson & Johnson and initially approved for emergency or conditional use in the United States, in the European Union, and elsewhere.

The technology for adenoviral vector vaccines is slightly more seasoned than the one for mRNA vaccines, as it was used for the Ebola vaccine, which was approved in 2019. However, the ability to produce billions of doses was only tested for the first time with the COVID-19 pandemic.

## Recombinant protein vaccines

Earlier we mentioned this type of vaccine, of which several have recently been developed against COVID-19. For instance, researchers at the US firm Novavax engineered a baculovirus, an insect-specific virus, by inserting the gene for the S glycoprotein into its genome. They then infected insect cells with this virus and grew them in culture. The infected cells were able to produce SARS-CoV-2 S proteins, which were then purified, assembled with nanoparticles, and ready for vaccine use.

Another protein-based vaccine is the one produced by the Russian Vector Institute, whose efficacy is still unclear.

This kind of vaccine takes slightly longer to develop when compared to fully or partially synthetic vaccines, because of the time needed to grow the modified baculovirus in cell culture. However, the data from the Novavax vaccine trial have been encouraging and it has been approved for emergency use in several countries, after the phase III trial with a measured efficacy of around 90 percent. This vaccine could offer more than one advantage.

Experience with this technology is vast, as it is used to produce licensed vaccines against influenza and HPV, while production is standardized. The requirements for distribution and storage are also simpler, as the vaccines remain stable in the refrigerator at 4°C for about three months.

Other vaccines of this kind that are currently being developed contain the S glycoprotein, which is produced in tobacco cells by the Canadian firm Medicago; in insect cells, by the French company Sanofi Pasteur; and in mammalian cells, by the South Korean firm SK Bioscience and by the Chinese company Clover. The GSK proprietary AS03 adjuvant, or the one made with synthetic nucleotides by Dynavax, has been added to these vaccines.

### Inactivated vaccines

These are the oldest vaccines for their design. We have already mentioned them in previous chapters. The few developed against COVID-19 contain the whole virus, obtained from patients and subsequently inactivated with chemicals. These kinds of vaccines were widely used in the past, for example, against polio and other diseases, and are founded on a vast experience. An adjuvant can be added to this type of vaccine so that the immune system reacts quickly and more effectively to the antigens. So far, inactivated vaccines, such as the Chinese Sinopharm and Sinovac, with respectively 79 percent and 50 percent efficacy, have been approved for emergency use in China and in other countries. However, the Sinovac vaccine has shown to have several limitations in the protection offered, particularly in countries very affected by numerous variants, such as Brazil.

## The efficacy measured in clinical trials and effectiveness in the real world

For most COVID-19 vaccines, it is important to note that the levels of efficacy were measured in clinical studies when the first version of SARS-CoV-2 virus was still prevalent. The variants that have since become dominant were either not yet in existence or in circulation by the time the experimentations were over.

As we already know from our experience with anti-flu vaccines, the effectiveness of each vaccine in the real world can decrease when it is administered to the population in post-trial periods, as the prevailing strands of the virus may have mutated slightly compared to earlier versions that were present only months beforehand.

Some vaccines against COVID-19 have shown a decreased effectiveness against some variants that have emerged or established themselves during the pandemic (as we will see later in this chapter). Second-generation vaccines are already being tested to cover against those variants that risk outwitting the vaccines currently being used.

## A race made possible by ...

When the pandemic hit, even the most experienced and optimistic scientists would not have bet money on the fact that safe, effective, and emergency-approved vaccines against COVID-19 would be available in less than a year.

Extraordinarily advanced research and technology, together with massive public funding and a full-blown pandemic, have helped to overtake even the most confident forecasts.

The speed of COVID-19 vaccine development has been truly amazing, as we can see from a graph comparing the previous amount of time needed to study and sometimes obtain vaccines against different diseases (Figure 3). How was it possible to obtain results in so short a time during this pandemic?

First of all, the genetic and molecular data for the virus and for the disease were published at the very beginning of the epidemic by Chinese scientists, which helped colleagues all over the world to work quickly and to share the results as they became available, in an equally open way, on web platforms accessible to all.

As all of the information about the SARS-CoV-2 virus became available, methods, technologies, and platforms were ready, thanks to all of the progress in vaccine research accumulated over a few decades. All of which would allow the entire scientific and biotechnological community, not just vaccinologists, to contribute to the development of vaccines. To give one example: Ugur Sahin and Özlem Türeci (Figure 4), the pair of researchers who developed the first mRNA vaccine that crossed the finish line, with the small company BioNTech in partnership with Pfizer, are a German couple of Turkish origin, and were absolutely unknown outside the small world of European biotechnology start-ups. Before COVID-19, they had been involved in cancer research.

Stéphan Bancel, one of the "fathers" of the second mRNA vaccine, produced by the US company Moderna, was slightly better known in the world of vaccinologists,

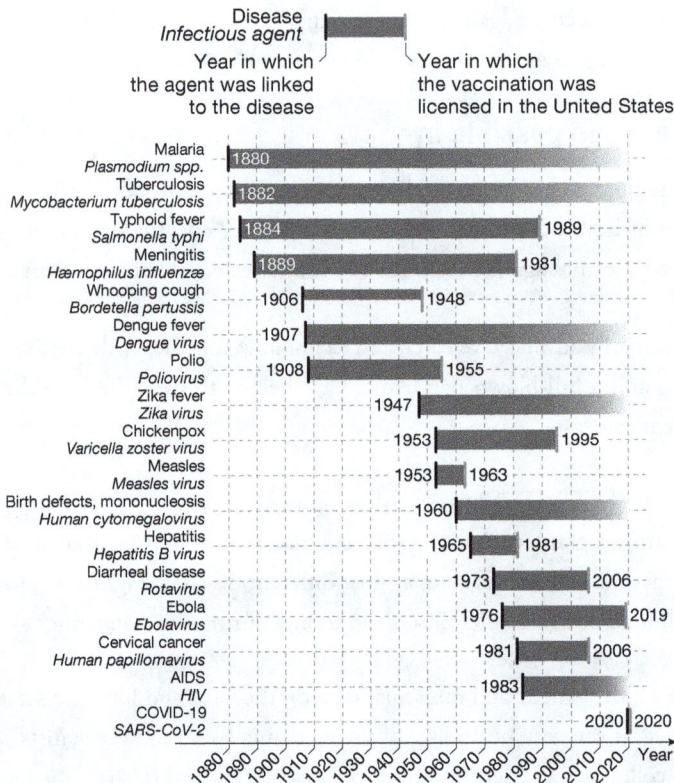

Figure 3: Timeline of innovation in the development of vaccines. Each bar begins in the year in which the pathogenic agent was first linked to the disease and the bar ends in the year in which a vaccination against that pathogen was licensed in the United States (Edouard Mathieu *et al.*, A global database of COVID-19 vaccinations, *Nature Human Behavior*, 10/5/2021).

Figure 4: Ugur Sahin and Özlem Türeci, the inventors of the BioNTech/Pfizer mRNA vaccine; Barney Graham and Jason McLellan, the academic "dads" of the mRNA vaccine produced by Moderna; Sarah Gilbert, at the origin of the adenoviral vector vaccine developed at the University of Oxford in collaboration with AstraZeneca.

but was far from being a world name. When Bancel left a larger company to become CEO of a small biotech in 2011, he warned his wife that the bet on mRNA-based technologies had a 5 percent chance of success. But he also told her that, should they win that bet, it would change the course of medicine.

More well-known in the world of vaccinology and coronaviruses are the already mentioned Americans: Barney Graham and Jason McLellan, and the English Sarah Gilbert, of the Jenner Institute of Oxford. The three had already worked on a candidate vaccine against MERS, the COVID-19-like disease caused by a previous coronavirus that emerged from dromedaries (we talked about it in Chapter 7).

## The best research products do not come out of nowhere

For some time, vaccinologists have been trying to develop vaccines based on nucleic acids, DNA and mRNA. Their reasoning was that these vaccines could instruct the cells of their human recipient to autonomously produce the viral antigen contained in the vaccine's instructions. As we have seen already from the first pioneering attempts by Craig Venter and Rino Rappuoli against influenza (see Chapter 6), the main advantages of these approaches lie in the time gained and in safety.

This is because in building this type of vaccine, all that is needed is the information on the viral genetic sequence, which can be shared online in a matter of seconds. There is no need to transport the virus, once isolated, to a laboratory, nor does it have to be propagated there in cultured cells, which is the case for more traditional vaccines. These steps require many months of work, especially if the virus is unknown, as was SARS-CoV-2, and when it is not initially clear if it can grow in cell culture. Also, when only information is involved, the need for viral propagation is removed, and so is the need for specially equipped and cumbersome laboratories for handling pathogenic microbes.

The other great novelty of mRNA vaccines is the chemical language used. The instructions for one of our cells to autonomously build a viral antigen must be written in an alphabet made of nucleotides, which is understandable to our cellular machines. But it must not be alarming to the cells of the immune system, who are always sensitive to any external agent. And so, the idea was born, in the minds of Kati Karikò and Drew Weissman, of producing an mRNA with nucleotides as similar to ours as possible, but different from those present in viruses and other microorganisms.

The mRNA produced in the nucleus of our cells undergoes modifications made by enzymes which, for example, introduce -CH2 groups, or isomerize uridine (U), transforming it into pseudouridine ($\psi$). Once modified, the mRNA is transported to the cytoplasm, where it associates with ribosomes and is used to synthesize proteins. If unmodified mRNA is found in the cytoplasm, this can only come from bacteria and viruses, and our cells react by producing inflammatory cytokines to stop the presumed infection. The modified mRNA, on the other hand, is not recognized by our innate immune system, and so it does not create alarm, nor cause any serious side effects.

The vaccinologists' idea was to produce mRNA with these characteristics, which would be safe for both administration and the rapid production of large quantities of stable mRNA. mRNA is a molecule that deteriorates quickly. For this reason, it is necessary to wrap it in the protective envelope of lipid nanoparticles that we mentioned earlier. The envelope also facilitates entry into the cell since the material is similar to that of the cell membrane with which it merges.

In the case of the adenoviral vector vaccines, the envelope that encloses the DNA sequence encoding for the S glycoprotein is a modified adenovirus. Decades of research have been dedicated to genetically modify viruses into harmless and efficient vehicles for transporting any substance into specific human cells, be it a vaccine or gene therapy.

Moreover, the advances in synthetic biology have been accompanied by those of reverse vaccinology, structural vaccinology, immunology, and monoclonal antibodies, that have assisted in the choice and development of the best possible antigen, as discussed in Chapter 4.

The most advanced vaccines against COVID-19 are therefore the poster child of all these advances in technologies and research methods, resulting from decades of intense work by the best minds in the field.

## The massive amount of money, mostly public, that has been invested in the development of the COVID-19 vaccines

All these advances and tools were available to the scientific community around the world at the beginning of 2020, through articles and publications that were freely accessible by anyone. But knowledge and technology will not go very far without money.

An immense amount of money has been invested in the most promising vaccine candidates that several companies around the world were developing in early 2020. The funds, largely public, came mainly from the United States, with more than $13 billion spent on Operation "Warp Speed," perhaps the best legacy of the otherwise woeful Trump administration. Other significant investments were made by the United Kingdom, followed by the European Union, China, and many other nations. To these, other funds were added, made available by international organizations such as the Coalition for Epidemic Preparedness and Innovation (CEPI), Gavi, the Vaccine Alliance, and the Bill & Melinda Gates Foundation (whose contribution to vaccinations for the poorest countries will be mentioned in Chapter 9). Even private citizens have offered their support. To cite just one example, the American singer Dolly Parton donated $1 million for the development of the Moderna vaccine.

This unprecedented investment has allowed companies to go through all phases of vaccine development in parallel as opposed to step by step, without any shortcuts. In nonemergency conditions, a company will generally complete one phase before starting the next. This makes sense, as a company cannot afford to build a production plant where a vaccine candidate, still in its development phase, has yet to be shown to work in clinical trials. Each phase is extremely expensive, and companies would go bankrupt if they went through all phases simultaneously.

For COVID-19 vaccines, the entrepreneurial risk, and the many inevitable failures of research and development, have been taken on almost entirely by the public sector, and in part by the nonprofit sector. Governments especially have provided companies with upfront funding, and with advance purchase orders for large quantities of vaccines.

A lot of money was therefore spent, right from the start, on setting up impressive production plants, in compliance with the very strict rules of GMP, and on prefilling hundreds of millions of vaccine vials. This was happening well before it would be known whether the vaccines would pass the tests for safety and efficacy and be deemed worthy of regulatory approval. But it was the only way to ensure that millions of vaccine doses would be ready for shipment on the very day the first companies received approval for emergency or conditional use.

Given the enormous amount of money invested, it is fair to ask: was it worth it? Will these investments ever pay off? Operation "Warp Speed"—the public-private partnership initiated by the US government to facilitate and accelerate the development, manufacturing, and distribution of COVID-19 vaccines, therapeutics, and diagnostics—has cost more than $13 billion. This may sound like a lot, but Michael Kremer, Professor at the University of Chicago and a Nobel Laureate in Economics in 2019, noted at the Trento Festival of Economics in June 2021, that the cost of the pandemic in the United States has been estimated (by David Cutler and Larry Summers) at $890 billion per month, which corresponds to about $30 billion per day. Thus, Operation "Warp Speed" would have paid for itself even if it could end the pandemic just twelve hours sooner. And this estimate only takes into account the gigantic economic losses, and not the human ones, caused by the virus.

## No shortcuts were taken in safety and efficacy tests (at least in countries where transparency is the rule)

The risks taken, which were enormous, were financial, and did not affect either safety or efficacy. Since safety and efficacy were such a big concern, each vaccine that reached the third phase of clinical trials, which is the largest and most expensive phase, could be evaluated on average on groups of 35,000 volunteers. Thanks to the large amount of funding available, this was 10–20 times the number of volunteers normally involved in nonemergency clinical trials.

In this way it was possible to evaluate both the efficacy of the preparations and the possible side effects that can occur in 1 out of 10,000–30,000 cases. For rarer effects, which occurred for instance in 1 case in 100,000 or 1 in 1 million when the authorized vaccines began to be administered to the population, pharmacovigilance (which we will be discussing in the next section) promptly set in.

The tremendous speed and extent of the pandemic was sadly (but happily for vaccines) assisted by the huge number of cases recorded in the viral waves that stormed over continents and hemispheres between the summer and autumn of 2020, when phase III clinical trials for the first candidate vaccines were carried out, leading then to the vaccines' approval.

As the Italian immunologist Guido Forni has explained, phase III is concluded when a sufficient number of cases of disease have been registered in the group of people recruited in the study, both in the vaccinated and in the control arm, to allow for a statistically significant analysis of the data. This kind of analysis has to clearly show whether the vaccine has a protective effect or not. Where there is no ongoing epidemic, it takes a long time to get a sufficiently large number of cases of disease in the group under observation. Instead, it was precisely the dramatic spread of COVID-19 that made it possible to quickly conclude the intermediate phase III of the study, which led to vaccine approval.

In conclusion, the vaccines' development and the clinical trials were conducted as fast as possible, given the available knowledge, technologies, funds, and the state of the pandemic. However, no shortcuts were taken on safety and efficacy, despite the strong political and economic pressures to run even faster. Had it been otherwise, the strict regulatory bodies in Western countries such as the

FDA and the EMA, would not have authorized the vaccines for emergency or conditional use.

All these guarantees apply to vaccines developed in countries with strict laws on drug testing and approval. To date, very little is known about the safety and efficacy of the Sputnik V adenoviral vector vaccine, which is based on two types of combined adenoviruses, and developed by the Russian research institute Gamaleya. In August 2020, President Putin announced the approval of this vaccine for emergency use before the phase III human trials had even begun. Sputnik V was then made available for vaccination campaigns in Russia and other countries, although the results of the clinical study were only published in scientific journals later, with numerous gaps. A similar lack of transparency on experimental data applies to some Chinese vaccines that have been administered in China and supplied to other countries.

## Postvaccination surveillance

By February, 2022, over ten billion vaccines have been administered worldwide since the start of the vaccination campaign. However, vaccination rates have been uneven around the world, with the majority so far concentrated in wealthier countries.

Among the billions of vaccinated people, most reactions after the shots had been already observed in clinical studies and were modest and tolerable: a few lines of fever, redness at the injection site, a headache or similar. Some have had more serious but treatable adverse reactions, such as allergies or, in some cases, anaphylactic shock. Anaphylaxis is a rare and severe allergic reaction to a certain substance, to an insect bite, certain foods, or drugs for instance. If detected quickly, it can be treated and resolved. This is precisely the reason why those who are administered a vaccine must wait 15–30 minutes after receiving the shot, so that this type of reaction can be treated immediately.

Very rare blood clotting disorders have been reported for vaccines with adenoviral vectors, such as those manufactured by AstraZeneca and Johnson & Johnson, which were not observed in clinical studies. The problem, more common in young adults and in females, is treatable if caught early. In fact, most of the patients were cured, but a few had a fatal outcome.

A problem that emerged with mRNA vaccines was myocarditis, an inflammation of the heart muscle, in about 2.7 cases per 100,000 vaccinated. Much more common following a SARS-CoV-2 infection, myocarditis typically resolves without consequences within a couple of weeks.

The timely reports and subsequent interventions by health authorities were made possible thanks to the activities of pharmacovigilance. As we saw in Chapter 3, pharmacovigilance closely follows any adverse reaction reported, at least in Western countries, and should an investigation be needed, administration of the vaccine can be temporarily suspended. After the most serious cases were reported and analyzed during the precautionary suspension of vaccinations in some countries, the health authorities asked the manufacturers to mention the possible rare side effects and their low probability as a warning on the package leaflet.

Given the extreme rarity of some of the most serious adverse events, the EMA, the FDA, and the WHO have all declared that the benefits offered by the adenoviral vector vaccines against the COVID-19 pandemic outweigh the risks, both for the individual and for society. The risks of serious or fatal adverse reactions are in fact limited, and almost always manageable. On the other hand, there is a very high risk of getting sick with COVID-19, a deadly disease for which there are very limited treatments. There is also a greater risk for society, should the pandemic continue beyond the bearable, given its serious social and economic consequences.

Nevertheless, some countries have restricted the use of this type of vaccine to older and lower-risk people, while other countries have abandoned them altogether. These decisions were not based on the mostly reassuring scientific evidence, especially in light of the risk-benefit ratio. What prevailed instead were the strong emotions provoked by the very rare occurrences, often accompanied by confusing communications from public health authorities and the media.

## How protected are we from SARS-CoV-2 and its variants after vaccinations?

In December 2020, the first vaccines approved for emergency use were finally giving a glimmer of hope for a possible exit from the pandemic, after almost a year of pitch darkness. However, in those very same days, some variants of the virus that had emerged in various parts of the world were being reported by health authorities and the media. Among the numerous and insignificant variants being signaled, a few were worrying the experts.

New variants emerge every day, given the virus' ability to frequently modify its RNA. However, only the so-called variants of concern have a greater capacity for contagion and for sowing disease and death. Since the beginning of the pandemic and until February 2022, the variants that have so far been considered worthy of attention and that have spread widely have been: "alpha," isolated in the United Kingdom; "beta" and "omicron," detected for the first time in South Africa; "gamma," identified in Brazil; and "delta," discovered in India.

In these variants, the genetic and protein structure is fundamentally unchanged from the originally isolated virus. However, in some cases changes can enable the virus to circumvent natural immunity developed after the disease, or to a lesser extent the immunity induced by vaccines.

Over time, new variants replace earlier viruses that have been blocked by immunity, whether induced by disease, or from vaccines.

The variants can also take root in populations that were heavily affected by earlier viruses, since natural immunity is insufficient to prevent new infections. The level of protection induced by neutralizing antibodies resulting from an initial infection is in fact significantly reduced against some viral variants. A tragic example is the Manaus region of Brazil, where the "gamma" variant has spread widely in the first half of 2021, despite 76 percent of the population being already seropositive for SARS-CoV-2.

There is good news, though. All vaccines approved so far in Western countries, if received in the recommended dosages and boosters, protect against serious illness, hospitalization, and death, which helps avoid the collapse of health systems. Furthermore, they act sufficiently against both the original virus and the variants that have emerged so far, up to February 2022. The kind of protection induced by vaccines is in fact superior to the one elicited by an infection, thanks to at least two components: more neutralizing antibodies of greater specificity, and T cells that are not particularly affected by variants.

The mRNA vaccines that show the maximum efficacy against symptomatic disease induce 3–4 times the concentration of neutralizing antibodies, or *titer*, found on average in the convalescent plasma of COVID-19 patients. Vaccines with adenoviral vectors lead to a titer approximately equal to that found in convalescent plasma, whereas the measured level of protection by inactivated virus vaccines, not approved in Western countries, is lower.

For all these reasons, it is essential that the majority of the population be vaccinated against COVID-19 with all the recommended dosages and boosters, including people that have already been infected and that, if unvaccinated, risk reinfection. It is also advisable to vaccinate younger people, as they can spread the infection to more fragile individuals, even if they are at lesser risk of falling seriously sick.

Should new variants lower the level of protection offered by vaccinations, steps are already being taken for second generation preparations, which will shortly be available for use. These could include protein subunit vaccines, which are more powerful than the ones which use mRNA.

Perhaps it will be necessary to subject at least the most vulnerable people to a seasonal vaccination campaign once a year, very much like what happens with the flu. Alternatively, a universal vaccine could be developed, against all types of coronaviruses that are the most threatening to humanity. For the time being though, these are just hypotheses, to which only time and a greater knowledge of the immunity generated by COVID-19 vaccines will give the most appropriate answers.

## How to put SARS-CoV-2 virus in jail

At this point, it should be clear that this new virus is playing a long tug of war with us. After it succeeded in confining us to our homes, we marked a point against the virus with the first vaccines. It then pulled the rope back with the variants of concern, to which we are now reacting by accelerating the vaccination campaign and preparing updated vaccines.

We could also see this fight as a plot for an action movie. Initially a character, the virus, bursts onto the scene to spread, unrestrained, all over the world. It avoids capture until it is finally placed in a prison, consisting of the natural immunity developed by those who have fallen ill. However, this prison is not sufficiently secure, and in fact, the virus can escape, especially if aided by some mutations. Vaccine-induced immunity is more like a maximum security prison, where escape routes are more tightly sealed. The film is not over yet, and plot twists are not excluded.

Given this scenario, we can keep the virus in prison by maintaining caution, protecting ourselves and others with masks, social distancing, and common sense. On the other hand, we can be confident that first and possibly second-generation vaccines will help close the gaps. Alongside prudent practices, monoclonal antibodies and antiviral therapeutics, approved or still in development, vaccines should be able to keep the current and future variants of SARS-CoV-2 under control, and progressively liberate us from the COVID-19 pandemic.

## Can people be vaccinated with doses of different vaccines?

The so-called "heterologous" vaccination, with two doses of different vaccines, is a strategy that has already been pursued, at least experimentally, with vaccines against influenza, Ebola, and HIV. In the case of COVID-19 vaccines, a first dose with an adenoviral vector vaccine could be followed by a second dose with an mRNA vaccine.

The idea behind the strategy is that different vaccines can show our defenses different parts of an antigen, as well as stimulate more components of the immune system, such as B and T cells, giving rise to a stronger protection. It can also assist in managing fluctuating supplies of vaccine doses more flexibly, or to replace one vaccine with another in case of problems.

However, the combination needs to be tested for both safety and efficacy. It could, in fact, give rise to undesirable reactions that require further study. Furthermore, it is necessary to test whether the combination works, given that in the past the mixed approach has at times failed (see the section on AIDS in Chapter 5).

Also, it is not easy to convince companies making different vaccines to collaborate in this kind of approach and with the clinical trials. However, the emergency situation and the strong public investments have pushed even the most reluctant to act responsibly and reasonably. As of February 2022, the heterologous vaccination studies undertaken against COVID-19 have produced encouraging results.

## How monoclonal antibodies contributed to the vaccines and to the fight against COVID-19

Antibodies are crucial elements of the immune system that are mobilized by every organism after encountering SARS-CoV-2, in an attempt to rid themselves of it. They are also essential in eliminating the virus in the course of the disease.

The antibodies developed by patients, and isolated by scientists from their plasma, have helped pinpoint the viral antigens that triggered the immune response during the COVID-19 disease. They have therefore been one of the most crucial elements for designing and optimizing vaccines and for evaluating their effectiveness. In this way, it was discovered that the virus was neutralized and eliminated mainly by the binding of one or more antibodies to the SARS-CoV-2 S1 protein.

Antibodies also tell us if a person has encountered SARS-CoV-2. Indeed, the presence of specific antibodies in serum, identified in the countless serological tests carried out on billions of people during the pandemic, is measured by reagents containing antibodies.

Thanks to monoclonal antibodies, we also know that the *antibody titer* after vaccination—that is, the quantity of neutralizing antibodies produced following two shots—is medium–high for mRNA vaccines, medium–low for those with an adenoviral vector, and very high for protein-based recombinants vaccines with adjuvant.

We must briefly mention the use of monoclonal antibodies for COVID-19 treatment. Following Emil von Behring's lesson on passive immunization (mentioned in Chapter 2), the so-called hyperimmune plasma obtained from recovering patients was used, especially at the beginning of the pandemic, as a possible therapy, albeit with mixed results for patients.

The modern and much more effective version of that treatment starts by discarding the many possibly useless or even dangerous parts of the plasma. Thanks to the extraordinary technological improvements of recent years, we now know how to identify the most effective antibodies present in the plasma of a convalescent patient and copy them, so to speak, in the lab, in the form of monoclonal antibodies. The power of these antibodies is up to one thousand times greater than what was obtained in the past, and this may help reduce dosages and costs, perhaps by a thousand to one. So far, because of their very high costs, monoclonal antibodies have only been used in the richest countries, in patients with serious oncologic and autoimmune diseases. In the COVID-19 pandemic, a tiny elite of high-risk patients, including former President Trump, were given these products.

Clinical studies have shown that this treatment avoids the need for hospitalization in about 75 percent of cases and speeds up the shedding of the virus. However, monoclonal antibodies for therapeutic use against viruses have also shown some limitations: they are effective only in the first few days of infection and not subsequently, when the disease has already been ongoing for some time and the inflammatory phase has taken over from the viral one.

Today, most patients are hospitalized in this second phase of the disease and in serious conditions, unresponsive to monoclonal antibody therapy.

In some cases, monoclonal antibodies can increase the severity of some infections, triggering major inflammatory reactions. However, this risk can be overcome by engineering the antibodies in the lab to eliminate the so-called Fc portion, the "foot" of the Y-shaped molecule, which can cause inflammation.

Another problem is that, by being very specific against certain parts of the virus, a monoclonal antibody may not work against a mutated virus. A possible solution is to use cocktails made with multiple antibodies to block all the variants together.

Monoclonal antibodies are therefore an extraordinary tool for developing vaccines and controlling their effectiveness. With all their limitations, they are also one of the few specific weapons we have for treating COVID-19 patients, along with antiviral drugs.

## The logistical challenges of vaccine production, distribution and mass administration

Many people have rightly said that the most innovative vaccines to emerge, in just eleven months of a frenzied coronavirus race, are a marvel of science. For these marvels of science to be useful as well as amazing, they must be produced in billions of doses, then distributed and administered to billions of people, despite the uncertainties engendered by a fluid and ever-changing situation. Vaccines should also be made available fairly and equally.

Considering that most vaccines require two doses and a booster, about 24 billion doses are needed to protect the world. A production capacity of these proportions was non-existent at the beginning of the pandemic and is still far from being achieved. Certainly, the situation has improved as the pandemic reaches its third year, thanks to massive funding, but the pace of production, however much speeded up, has struggled to keep up with the much faster rate of contagion. Much of what is needed to produce COVID-19 vaccines are in short supply.

The delays in delivery, observed mainly at the beginning of the vaccination campaign, have mostly been due to companies struggling to secure high quality reagents and chemicals, as well as highly specialized personnel, fast enough.

Other challenges include: ensuring the supply of vials and caps in which to insert the vaccine; acquiring unprecedented numbers of liquid nitrogen tanks, −80°C freezers (and the necessary electrical power), possibly worldwide; producing enough dry ice in which to store the doses of the most perishable vaccines; maintaining the cold chain; disposing of enough airplanes and trucks for transport; guaranteeing the physical and digital security of vaccines, which at the peak of the crisis were tantamount to "liquid gold"; transforming stadiums, gyms, and other large venues into temporary vaccination hubs; establishing the order of priority for administering vaccines to the different population groups (for instance, medical and healthcare personnel; law enforcement officers; nursing home staff and guests; the remaining sections of the population, taking into account age and health status, and so on); building a safe and effective IT infrastructure to track vaccine recipients, vaccination dates and potential adverse reactions; and hiring and training temporary staff.

This long list, which is just a sample of the many complicated activities needed for the extremely complex COVID-19 vaccination campaign, should give us pause for thought. Instead of marveling at how much was done quickly and done well, we paid more attention to what, inevitably, went a little wrong, and rushed to criticize and argue on social media networks, without even acknowledging that the errors that happen are often corrected quickly.

Sometimes, citizens of wealthier countries fail to realize how, even in these current circumstances, they are far, far luckier than the majority of people in the world living in poor countries. Based on confirmed vaccine orders, as of January 19, 2021, the highest-income countries had already secured, at least on paper, 5.3 billion doses through advance order contracts, compared to just 681 million doses for lower income-countries (Figure 5).

Updated: January 19, 2021

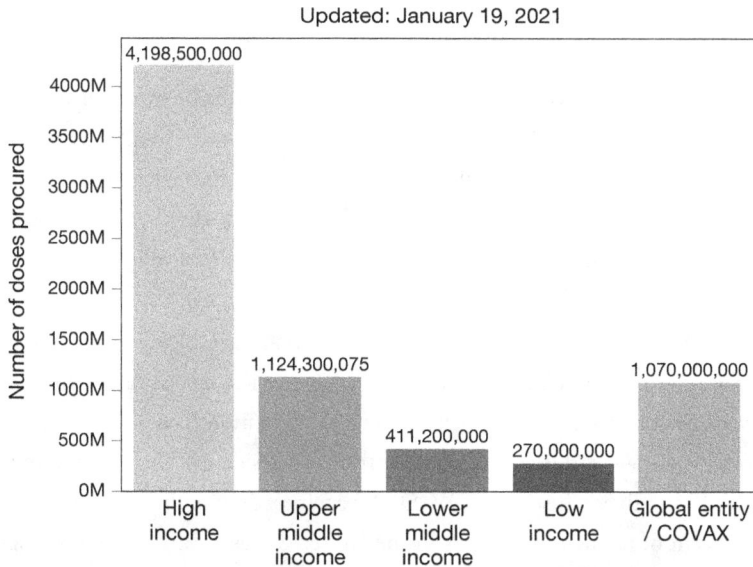

Figure 5: The vaccine doses ordered by the different countries by income level, as of January 21, 2021 (Rino Rappuoli).

Even among the wealthiest nations, the immense challenge of this unprecedented vaccination campaign has created multiple tensions. For instance, on January 2021 the European Union and the United Kingdom had a serious argument over the then still scarce doses of the AstraZeneca vaccine, in the shadow of the bitter "divorce" over Brexit.

## How to convince the hesitant?

A not insignificant part of the world's population has expressed doubts about the idea of getting vaccinated against COVID-19 to protect themselves and the people around them. Their doubts have been circulating loudly and widely online and on social media, despite the almost six million deaths recorded as of February 2022, and the immense efforts to obtain and distribute effective and safe vaccines in record time against the horrible COVID-19 disease.

In Chapter 3 we saw the many reasons, mostly irrational and none based on science, that are often adduced by people who distrust vaccines. We have also seen how this state of confusion is often fomented by entities with vested interests, intent on sowing uncertainty and distrust. If these reasons are disconcerting in times when microbial threats have been kept very well at bay by vaccines, they are truly incomprehensible in the midst of a pandemic that has already taken the lives of millions. It is even more inexplicable, since it is becoming increasingly clear that the pandemic's black cloud can only be dispelled by mass vaccinations reaching a coverage of at least 85–90 percent of the population to achieve a vast immunity.

At the time of the polio epidemic, in the 1950s, the arrival of the first Salk vaccine was greeted by church bells ringing in celebration. Today, a considerable part of the staff working in nursing homes and residences for the elderly has declared that they refuse the COVID-19 vaccines, despite the fact that around 40 percent of pandemic's victims in the United States in 2020 were the fragile guests of those institutions. As a consequence, the COVID-19 vaccines for healthcare professionals have now been made mandatory in several countries.

As Jonas Salk's son Darrell said in an interview with *The New York Times*, his father "would have been delighted" with the vaccines being made available in such a short time, "but he would be horrified by the number of people concerned about using the vaccine. I can see him closing his eyes and shaking his head."

Yet, there is cause for hope. Vaccine acceptance began to increase when, from being a mere hypothesis, COVID-19 vaccines became a reality. The clinical trials data, showing efficacy up to 95 percent for some vaccines and modest side effects, convinced many previously undecided people, and also because these results were published just as the pandemic was reaching frightening peaks of new cases and deaths. In some New York hospitals there was even competition and rivalry among staff for the first few doses supplied. The moment also coincided with the transition from the Trump administration to that guided by President Joe Biden. The new President has helped make the public narrative about the virus and the pandemic in the United States more factual, with important repercussions for the rest of the world.

To further convince the still undecided, additional motivation could come from associated benefits, such as being allowed to travel without restrictions, being admitted to a football stadium, or to a crowded concert. In short, the vaccinated "class" may have some advantages in resuming a more normal life earlier.

## Will it be possible to vaccinate an entire planet against COVID-19?

The numbers updated to February 2022 tell us that Latin America and the Caribbean countries have suffered about a quarter of the world deaths for COVID-19, while Africa has exceeded 11 million cases and India has recorded more than 42 million cases and over 509,000 deaths. Quarantines and lockdowns have hit the poorest countries even harder, where homes are often overcrowded, remote work is virtually impossible, no safety nets or subsidies are available for those who have lost their jobs, and the vaccination campaigns are difficult to implement because of lack of doses, poor health facilities, and large populations.

Since the beginning of the pandemic, attempts have been made to develop and produce COVID-19 vaccines for the more than 150 countries that cannot afford them. As we will also see in the next chapter, this effort, called COVAX, was coordinated by the WHO together with the Bill & Melinda Gates Foundation and other entities, such as the CEPI and Gavi, the Vaccine Alliance. The European Union, China, and many other nations have contributed to varying degrees to COVAX. On the other hand, during President Trump's administration, the United States refused to participate to this effort. In July 2020, the Trump administration had also withdrawn the United States from the WHO, which consequently was missing more than $400 million from annual American contributions. Fortunately, the Biden administration reversed these decisions and offered to donate a few hundred million doses of vaccines.

COVAX has mostly pointed to the cheaper and easier vaccines in terms of storage and transport. The University of Oxford, in partnership with AstraZeneca, committed to offering nonexclusive, royalty-free licenses on its vaccine to manufacturers in poor countries. The Serum Institute of India had agreed to produce it, even though the very serious Indian pandemic crisis has hampered the work in the plants. Other companies involved in COVAX were Novavax, Sanofi, and Johnson & Johnson. Another promising vaccine for poor countries is Corbevax. Developed at the Baylor College of Medicine, in Houston, Texas, by Peter Hotez and Maria Elena Bottazzi, Corbevax is a protein vaccine that has been proved safe and effective in clinical trials and it has been approved for emergency use in India at the end of 2021. The mRNA vaccines have instead been used so far almost exclusively in wealthier countries.

However, financial efforts so far are still largely insufficient for reaching the 24 billion doses needed to vaccinate the entire planet, and this is a great injustice. According to estimates made at the end of 2020, poor countries would receive enough doses to inoculate just 20 percent of their populations by the end of 2021, while even more pessimistic forecasts show that there will not be enough vaccines to cover the entire world until 2024.

The duration of the pandemic, the uncertainty about the future, and the feeling that citizens of poor countries were not considered enough in the battle against COVID-19, have fueled resentment and conspiracy theories. Travelling on the web, they can undermine vaccination programs worldwide. Among the most fanciful inventions, an idea was widely circulating that Bill Gates's twenty-year commitment to vaccinations in the poorest countries was in fact a cover for implanting microchips and tracking vaccinated people. Obviously, it was a bogus, unsubstantiated theory.

The populations of poor countries must be included in the effort to free the world from the scourge of COVID-19 if we want the planet to return to being a safe and secure place from the virus, and if we want economies and societies to fully recover. Likewise, in the interest of public health, vaccines must be offered to irregular immigrants in rich countries as well.

For the future, there are proposals to equip Africa with a greater capacity to produce its own vaccines. This idea is commendable. However, a state-of-the-art production capacity for emergency vaccines requires huge investments in plants, raw materials, and personnel. All these resources must be kept in place, updated, maintained, and paid for for all the decades that may pass between one pandemic and the next. Once the emergency is over, there is a risk that any investments would be terminated quickly without the commitment and support from the public sector and the international community.

## Would a patent waiver on vaccines really accelerate the rate of immunization in the whole world?

As of February 2022, a huge, unfair gap separates the vaccinated people in the richest countries, where at least 70 percent of the population have been immunized, from the poorest countries, where shots have been given to just 10 percent of the population so far.

Some politicians have expressed the opinion, shared by a part of the public, that a temporary patent waiver on COVID-19 vaccines would make it quicker and easier to vaccinate the whole world against the disease. The idea seems to be reinforced by the fact that, given the state of emergency, the development costs of most vaccines are largely sustained by public funds.

A patent is a type of intellectual property that gives its owner the legal right to exclude others from making, using, or selling an invention, for a limited period of time, when the owner retains the priority and the exclusive right of industrial exploitation.

Bringing a vaccine from an idea to the market can cost a company as much as $1 billion. No company would contemplate this sort of investment without the protection of its intellectual property through a patent as a guarantee for recovering costs incurred through revenues over a reasonable period of time.

A patent waiver, even temporary, may not end up being the most practical and effective solution to the problems of equity and access to vaccines. Actually, it could even be rather counterproductive, as Tony Fauci declared in May 2021.

If this pandemic were to go on (hopefully not!), or a new pandemic were to hit in the future, companies may be unwilling to develop new vaccines without the protection afforded by patents.

Moreover, for highly technological vaccines such as those developed against COVID-19, the manufacturing process is far from easy, even when licenses are offered at no cost. This is the case for the Moderna mRNA vaccine. In fact, the know-how, raw materials, technological platform, and highly specialized personnel needed to produce these vaccines are all very expensive ingredients, and are not readily available. In fact, the greatest obstacle to the production of vaccines in the poorest countries was precisely the absence of plants and the lack of trained personnel.

A more concrete and pragmatic option is to negotiate big discounts on prices, via a collaboration between the public and private sectors. Substantial reductions can be obtained from companies producing the vaccines, at the request of governments ordering and acquiring stock. This always holds true, but even more so during an emergency situation, when the vaccination of a high percentage of the population is a must. Also, rich countries can buy vaccines not only for their own citizens, but also for countries that cannot afford them, at lower prices. In this way, they would contribute to a more equitable right to health for the entire world population, as well as a faster exit from the pandemic.

## Communication in the time of COVID-19

The level of detailed scrutiny accompanying the COVID-19 pandemic, moment by moment, and molecule by molecule, is unprecedented in human history. From the variety of symptoms and different diseases that the virus has caused in patients, to the pandemic consequences for global health systems, economics, politics, social and public life, every single piece of news was poured out into the media, immediately, incessantly, unfiltered.

On the one hand, there is no doubt that every little detail of this relentless inquiry has helped doctors and scientists tremendously in understanding the many aspects of an unknown disease and for a prompt development of the vaccines that were desperately needed. On the other hand, this deluge of information, unmanageable even for the most techno-savvy and experienced, has been pouring out in full and in real time onto the public, mostly on social media and television.

People, frightened and disoriented, have tried to protect themselves as best as they could, not only from the virus, but also from this never-ending information flow. From this morass of information, everyone has grasped, at any given moment, what was most comfortable or least displeasing. What has been retained or discarded has a lot to say about the psychology of each individual, their intellectual skills, and level of education. The way people behaved during the pandemic, responsibly or recklessly, was possibly a consequence as well.

Perhaps the greater novelty, at least in the history of epidemics, was the fact that every citizen was able to publicly react to this incessant flow of data and hypotheses, proposals, and rules, and in turn spread their own moods and opinions. Emotional outbursts have in this way fueled an out of control, endless chain of externalizations, replete with unverified data.

Within this confused and complicated scenario, for anyone bearing authority and responsibility towards the public—whether scientist, politician, or journalist—it has been particularly difficult to inform people of what was taking place. Providing advice as to the most appropriate behavior to be adopted at any given moment has also been complicated. In the meantime, the virus roared ahead, upsetting life as we knew it, virtually everywhere around the globe, always a few steps ahead of the best attempts at prevention, containment, cure, and communication.

Any prediction as to what would happen the next day, in a week or a month, was promptly disproved by the unpredictable trend of the pandemic and by its social, economic, and human consequences.

What alternatives were there? Omitting parts of the truth, or lying altogether, were terrible choices for many politicians and authorities, who quickly lost any vestige of credibility before a frightened and disoriented public. Saying nothing? This is no longer an option in the age of internet and social media, where people will inevitably access countless fragments of information, most of them unreliable or distorted.

The ones that fared best seem to be the very few politicians, authorities, and journalists that said, honestly and clearly, what they did know, what they did not know, and what they did know up to a point. In this, they had to act like acrobats, seeking a continuously precarious balance on a tightrope. A tightrope that could easily lead the reader, or listener, to fall into an abyss of paralyzing fear or, vice versa, into an excess of hope and optimism.

Optimism is a feeling that, in serious situations like a pandemic, must certainly be nurtured. However, prematurely portraying a rosy picture can be risky, as people may consider themselves out of danger too soon, and so abandon any protection strategy. Conversely, stepping too hard on the fear pedal can easily result in an excessive level of anxiety and bad news for people to endure, with a sense of profound discouragement and mistrust setting in, giving way to a very dangerous indifference to the risk of contagion for oneself and for others.

The partial and provisional lesson, here, is that the microorganisms that threaten our health catch us off guard at speeds too fast for the mind's average ability of assimilation and adaptation. This is partly why we react in unpredictable ways, at times even against our best interests.

Whoever has the duty of guiding, governing, and providing forecasts in this situation is dealing with an art: the art of health communication in the days of social media, which is imperfect and rapidly changing. To this, add a level of uncertainty so high that even the most honest and capable communicator is constantly juggling between being a little bit right and a little bit wrong.

Many professionals, in good faith and to the best of their ability, have at least tried to speak and act sensibly, as best as could be done under the circumstances. They have tried to inform on the basis of the provisional results of science, in a continually changing picture, even at the cost of incurring the resentment of the many, many people that were confused, disheartened, or desperate.

# 9
# Vaccines for Whom?

What kind of vaccines will we build tomorrow? The future of vaccines depends a lot on the knowledge and technologies that we will be able to develop. Above all, it will depend on who will be the main "customers" of this newly applied knowledge. To use the language of economics, the future will be shaped by demand.

So far, the people who have benefited the most from vaccines are the inhabitants of the wealthiest and most advanced nations. In the future, vaccines promise to make their lives even more enjoyable and long-lasting, if we can tackle some challenges.

## Will we win antimicrobial resistance?

"Do you have an infection? Take an antibiotic." This very simple solution has saved millions of human lives in over seventy years since the introduction of penicillin, the first antibacterial drug.

However, widespread use of antibiotics has provided bacteria with an extraordinary evolutionary opportunity to beat our drugs. Out of the entire population of bacteria colonizing a tissue or an organ, it is possible for some of them to mutate. If the mutation helps the bacteria that carry it to dodge antibiotics, they will continue to reproduce and spread, while the nonmutated bacteria will be exterminated.

In fact, antibiotics act as a formidable sieve on that "lucky" population of bacteria that reproduce and proliferate over time (their luck, not ours!). Much the same mechanism favors the emergence of other types of microbes, in addition to bacteria, that are drug resistant.

An increasing number of infections, such as pneumonia, TB, gonorrhea, and salmonellosis, are becoming increasingly difficult to cure, as the antibiotics in use have become less and less effective. In the poorest countries especially, a lack of clean water and sanitation have compounded the problem, facilitating the spread of all microbes, resistant ones included.

According to some estimates, deaths from antimicrobial resistance (AMR) in the world are at least 700,000 per year, with the WHO naming AMR as a global threat to health and development, and one of the ten greatest challenges to world public health.

The main driver behind the development of drug-resistant pathogens is the misuse, or excessive use, of antimicrobials in both humans and animals. For example, a person who has come down with the flu may take an antibiotic for it—even if antibiotics have no effect on the flu virus—a common occurrence, as it is not always easy to diagnose whether a disease with similar symptoms is caused by a virus or by bacteria.

As for animals, antibiotics are widely and improperly used on farms, whereby preventing infections they promote a faster growth of livestock, ensuring lower costs and higher profits.

We have been aware of this issue since at least the 1950s. Stanley Falkow, the late professor of microbiology and immunology at Stanford University, in California, was among the first to understand how resistance can spread very quickly in a population of bacteria. In 1964, Falkow had isolated a gene capable of conferring resistance to antibiotics, in a portion of bacterial DNA called a plasmid. As early as 1975, he had observed how the very success of antibiotics in reducing mortality from bacterial infections, and their consequently widespread use, was at the origin of the emerging problems of resistance.

The costs of AMR to society and individuals are considerable: in addition to an increase in overall mortality, it also causes longer hospital stays, higher medical costs, and economic hardship for those affected. A UK report has warned that unless the situation changes, AMR could cause up to ten million deaths a year, surpassing cancer deaths, and causing economic damage in the hundreds of trillions of dollars by 2050. According to another model developed by the World Bank, the global economy could lose to AMR up to 3.8 percent of annual gross domestic product by 2050.

Today, many microbes show resistance to more than one drug, making the treatment of various infections more difficult. Of these, the most notorious and fearsome resistant bacteria are *Staphylococcus aureus* and *Clostridium difficile*. Cases of drug resistance are also emerging in patients on long-term treatment with antivirals, such as those against HIV. One of the best-known malaria therapies, a combination of artemisinin and other drugs, is partially ineffective due to resistance that has developed in several Asian countries. Infections caused by fungi, such as *Candida auris*, have also stopped responding to some types of antifungal drugs.

Even a trivial tooth extraction could become a life-threatening surgery, and the achievements of modern medicine that we take so much for granted risk turning into blunted weapons, causing the most complex surgeries to become unfeasible. If new antimicrobials are not developed against the most resistant pathogens, the most fragile patients, such as those undergoing anticancer chemotherapy, are at a high risk of mortality because they are particularly sensitive to infections.

While the development of resistance is almost inevitable with any antibiotic, the phenomenon is rare in the case of vaccines. There are several reasons for this. First of all, vaccines are mainly used in a preventive manner, when the populations of a pathogen in circulation are relatively small. This statistically reduces the probability that microbes will appear and proliferate, along with mutations that are capable of conferring resistance. Secondly, many vaccines are able to intercept microbes with multi-pronged attacks, which means that the pathogen would have to undergo multiple mutations simultaneously to become fully resistant.

Fortunately, with the latest technologies, it is not impossible to imagine vaccines for at least some of these microbes, and several hypotheses are currently under study. The same cannot be said for antibiotics, which have not seen any major innovations in recent decades.

One of the vaccines already developed and in use against resistant microbes is the one conjugated against *pneumococcus*. Data for use of this vaccine in Africa, the United States, and Europe have shown a reduction in the number of cases of antibiotic-resistant disease and a decline in the circulation of the resistant strains.

It has also been shown that people vaccinated against seasonal flu consume about 60 percent fewer antibiotics than unvaccinated people. Vaccinations therefore appear to have a significant direct and indirect impact on drug resistance.

A further but no less important reason for favoring vaccines over antibiotics is that, in this way, the microbiota is preserved. The microbiota is the set of microorganisms that we host in our body which are essential for health and growth. Our bodies are hotels for microbes: according to some estimates, one in two cells in our body does not belong to us. A course of antibiotics indiscriminately kills all bacteria in the body that are sensitive to the active ingredient, not just the pathogens responsible for the infection. Resistant bacteria grow more easily in a damaged microbiota, with the risk of harmful repercussions on health. In children especially, an altered microbiota can impair the immune system's development and affect nutrition.

The microbes that accompany us throughout our lives are vital for human health, including the metabolites they produce that are essential for the health of our organism. We now know that the human immune system functions poorly in the absence of bacteria in the intestine. The results of a study conducted by Bali Pulendran, of the Emory Vaccine Center in Atlanta, Georgia, showed that the flu vaccine is not effective in mice treated with antibiotics that kill the intestinal flora. This data, amongst others, suggests that our immune system's ability to fight infections, as well as prevent and eliminate tumors, is very much conditioned by the assistance it receives from our human microbiome.

## Vaccines for the elderly

Other challenges for vaccinologists serving wealthier (and older) clients are cancer and other chronic diseases—the ills of an aging society.

Therapeutic cancer vaccines are a type of treatment that can help kill cancer cells by stimulating the body's natural defenses. Unlike the vaccines that we have mostly dealt with in this book, therapeutic cancer vaccines are not preventive but are designed for people who already have cancer.

The idea behind therapeutic vaccines is that cancer cells expose proteins on their surface that are different or more numerous than those found on healthy cells. Theoretically, these features should allow them to be recognized and destroyed by the immune system. But this mechanism does not always work thoroughly and quickly enough to eliminate aggressive and rapidly dividing cells, like the ones that make up cancer. Cancer also tends to inhibit the immune system. Therapeutic vaccines can help a somewhat sleepy immune system, affected by the tumor, to recognize those antigens more effectively, quickly destroying all cancer cells with exposed surface antigens.

So far, the development of this type of vaccine has been rather slow, mainly because it is not easy to identify targets that are, if not exclusive, at least very specific for cancer. To date, only one therapeutic vaccine of this kind has been approved: called sipuleucel-T, it is indicated for advanced prostate cancer that does not respond to hormone treatments. Further developments in immunology and structural biology are opening promising horizons for other types of cancer vaccines.

Another vaccine that can help aging people is the one for *Herpes zoster*. This is the virus that causes chickenpox in children, but in older adults it provokes shingles: this happens when the virus reawakens, after decades of lying dormant in nerve endings following a bout of chickenpox. It is unclear why the virus is reactivated; a weakened immune system, or stimulation from a bout of chickenpox in a grandchild, for instance, could be plausible mechanisms. When it does flare up, shingles can cause very painful skin rashes and possible serious consequences, mostly for the eyes and the nervous system. A recombinant vaccine that contains the *Herpes zoster* glycoprotein E with an effective adjuvant is now available against this virus, which often causes relapses. Approved by the FDA in 2017 and by EMA in 2018, the vaccine is recommended for adults over fifty who both have or have not had episodes of the disease. The vaccine can prevent both zoster's acute manifestations and postherpetic neuralgia, one of the most debilitating aftereffects of this chronic infection.

## More affordable vaccines for poor countries

How much does a vaccine cost? In the pricelist available on the US CDC website, updated to February 2022, a dose of the cheapest vaccine costs about $9. However, among the most recent and advanced vaccines, several exceed $190 per vial. And these are prices for public health institutions, the costs are far higher when the customer is a private individual.

These prices barely reflect the huge costs that a company must sustain to create a vaccine, from its conception to its production. But the price also reflects demand: a vaccine that is used around the world will cost far less than a preparation that is useful to only a small section of population.

For instance, earlier in this book we referred to the recurring outbreaks of meningitis in the United Kingdom and New Zealand, which stopped after the populations of both countries were immunized with two specially prepared vaccines.

The investment needed to develop the vaccine for the more than sixty million inhabitants in the United Kingdom was barely profitable for Chiron, its manufacturer. The optimal collaboration with the British government, which provided the vaccine to the population as soon as it came out of the labs, limited the financial risks.

Whereas in the case of New Zealand, no company would have ever invested in a vaccine for a population that at the time ranged around the four million mark—just over the population of a city like Hanoi, in Vietnam—had the country's government not provided the $200 million necessary for its development.

All is well, so to speak, as long as infections strike countries that can afford the cost of developing a vaccine on their own. But what happens in the case of those nations that cannot afford it? The people get sick and die.

The region of sub-Saharan Africa, tragically known as the *meningitis belt*, ranges from Senegal to Ethiopia. Here we have almost yearly meningitis epidemics, with about thirty thousand cases, adding to an already long list of other diseases and problems.

Fortunately, even in these very poor areas, we are seeing improvements that give us room for hope. A few, quite specific vaccines have been available for some time, and they have been used both for prevention and in response to new outbreaks. But the biggest news is a conjugate vaccine against meningococcal A (Men A) meningitis that became available in 2010. Since then, It has been used in preventive immunization campaigns across the region which, by 2017, had involved more than 280 million people in twenty-one countries, with children now routinely immunized with this vaccine. The results are impressive: meningitis epidemics caused by meningococcus A have fallen by about 60 percent in the meningitis belt.

Economists maintain, data in hand, that a nation's development is closely linked to the general health of its inhabitants. They also say that vaccines are the most effective way to "buy" health and prosperity. Estimates for the ninety-four lowest-income countries globally show that for every single dollar invested in vaccinations, there would be a saving of $16 in healthcare costs, wages, and productivity losses resulting from disease and death. These figures, from a 2016 study by researchers of the Johns Hopkins University in Baltimore, Maryland, also show that if you include broader benefits, such as the value people place on healthier, longer lives, and the long-term burdens of disabilities, the net return increases to $44 per dollar invested. According to David Bloom, Professor of Economics and Demography at Harvard University, "the return on investment for many vaccines appears, conservatively, to be at least as high as the return on investment for spending on primary education." But it is precisely the lack of money that often prevents that initial investment, condemning countries with already poor and sick populations to conditions of ever greater infirmity and poverty.

The problem is that humanitarian needs do not go hand in hand with those of profit. Developing a vaccine from scratch costs an average of $500–$1,000 million. If a country is too poor, the gap between the amount the company needs to invest and the return that it can expect from vaccine sales is too wide.

This creates a vicious circle from which it is difficult to escape: faced with an uncertain demand for vaccines from poor countries, companies respond with a limited supply and high prices.

## The great alliance

We have had vaccines now for a few hundred years, but the possibility for children to access these preparations varies greatly depending on where they are lucky (or unlucky) enough to be born. By the late 1990s, progress of international vaccination programs in poorer countries had essentially stalled. For about thirty million children, vaccination schedules were incomplete, while many others were not vaccinated at all, with resulting exposure to many contagious and deadly diseases. At the same time, newer, safer, and more effective vaccines were becoming available, but countries where children were less protected, and where the need was the greatest, were unable to afford even the oldest and cheapest vaccines.

The situation has improved since then: in 1990, the age group with the highest mortality rate was by far that of children under five, however, statistics for 2017 tell us that deaths among the youngest have halved and that the age group with the highest mortality rate has now shifted to 80–84. This means that, at least

Million of deaths by age group worldwide

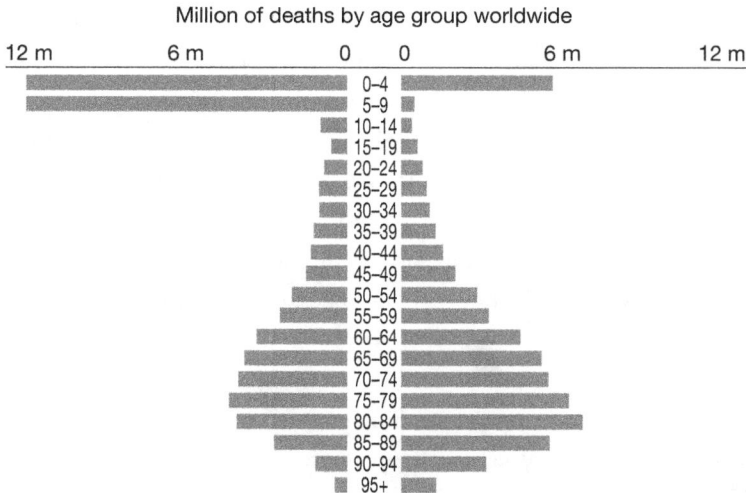

Figure 1: On the left, the 1990 data shows that the highest mortality in the world was by far that of children by the age of five; on the right, the 2017 data tells us that in the world mainly people aged 80 to 84 were dying, a sign of the progress made against infectious diseases of childhood (adapted from *Gates Notes; IHMI*).

until COVID-19, more and more people were coming to enjoy a longer life with children and grandchildren, thanks to the prevention of childhood diseases around the world (Figure 1).

In 2000, the Bill & Melinda Gates Foundation, together with United Nations Children's Fund, the WHO, and the World Bank, contributed significantly to a radical change by creating what is now called "Gavi, the Vaccine Alliance." It is an international public, private, and nonprofit partnership aimed at creating equal access to new or underused vaccines for children living in the poorest countries, reducing poverty, and protecting the world from the threat of epidemics.

In addition to its founders, the alliance involves civic organizations, public health institutions, governments of donor and recipient countries, as well as representatives of vaccine manufacturers.

What does Gavi do in practice? Firstly, it selects countries where immunization rates are lowest and where large numbers of children are not vaccinated.

Today, there are seventy-three selected countries and Gavi works with each of their respective governments. They aim to increase the number of children vaccinated; at improving the technologies and vaccines used (at the moment Gavi is focusing its efforts on seventeen infectious diseases); and at strengthening the capacities of each national health system to provide vaccinations in a sustainable way over time. The goal is to ensure that each country supported by Gavi can maintain its own vaccination system, gradually becoming independent from international aid. At the beginning of 2020, sixteen countries had already succeeded in this feat.

Gavi pools the demand for vaccines from poorer countries so as to access lower prices from producers. The nations represented by Gavi are where more than half of global births happen, which means they represent a large and vital market for vaccine manufacturers who, stimulated by an increased demand, can reduce their prices.

All the countries involved contribute towards the cost of the vaccines obtained through Gavi, though the quotas varying according to each individual country's ability to pay. As a country's income grows, the quota gradually increases until it can fully cover the vaccines' costs.

What has Gavi achieved so far? Since its establishment, Gavi has contributed to the routine vaccination of around half of the world's children, an impressive 888 million, preventing over 15 million deaths. It has also helped strengthen health systems and immunization services in 69 countries. By preventing long-term illness, death, and disability, between 2000 and 2017 vaccinations supported by Gavi are estimated to have helped generate economic benefits of over $150 billion.

Here are some of the Alliance's greatest successes (we cannot name them all, even if they all deserve it). Over 81 percent of children in Gavi-supported countries have been vaccinated with the pentavalent vaccine that protects against five lethal childhood diseases: diphtheria, tetanus, whooping cough, hepatitis B, and type b *hæmophilus*. A single dose of this vaccine used to cost

$3.65, but today, thanks to Gavi, it costs less than $1. The anti-HBV vaccine that is now included in the pentavalent had existed as a single vaccine for over 40 years, but for the 73 countries where Gavi operates, in 2000 vaccination coverage was less than 20 percent.

Other goals achieved were the more than 215 million children vaccinated against pneumococcal diseases and 332 million people involved in the campaign against meningococcal meningitis A in Africa, more precisely, in the meningitis belt we spoke about earlier. Gavi also contributes to the purchase of vaccine stocks to control any new outbreaks for these types of recurring diseases. In addition, more than 125 million children have been vaccinated against rotavirus, and, by 2019, more than 4.8 million girls have been protected against HPV.

The HPV vaccine prevents infection, mainly transmitted through sexual contact, associated with the development of cervical cancer, as well as cancer of the anus, vulva, vagina, penis, and oropharynx. The first anti-HPV vaccine, marketed by Merck laboratories, was approved in 2007 in over eighty countries, and it was then followed by additional vaccines against other cancer-causing HPV strains. Of the different types of cancer affected, cervical cancer is the most frequent, killing over 300,000 young women a year, especially in poorer countries where *Pap smears*[1] are not common practice for early diagnosis of the disease. Thanks also to Gavi, women in these countries would now stop dying of cervical cancer, instead of waiting decades before they could benefit from this excellent, effective, and safe (but expensive) invention of advanced research.

Another of Gavi's important contributions is the purchase of Ebola vaccines, approved in 2019. Ebola is a viral hemorrhagic disease that sometimes gives rise to epidemics in sub-Saharan Africa, where the virus is more common. In those regions, it is believed that the virus spreads at low speed among some

---

[1] *Pap* is an abbreviated reference to Georgios Papanicolaou, not to papillomavirus. Papanicolaou was a Greek doctor who, in the twentieth century, invented a simple procedure where a small brush or spatula is used to gently remove cells from the cervix. The cells are then examined under a microscope for cervical cancer or cell changes that can lead to cervical cancer. The pap test has saved millions of women lives from cervical cancer.

animal populations. People occasionally become ill after coming into contact with infected animals, and this can then result in epidemics that are spread from person to person. In Western countries, Ebola cases have been very rare and exclusively imported from abroad. Particularly worrying were the African Ebola epidemics that started in the second decade of this century and led to the rapid development of the first viral-vector vaccine, approved in 2019. The vaccine was created by Canadian researchers, with the gene of a viral glycoprotein, inserted by genetic engineering into a modified vesicular stomatitis virus. The Italian vaccinologist Riccardo Cortese gave a crucial contribution to the design of this type of vaccine.

Gavi is also heavily involved in the response to the COVID-19 emergency. Providing instant funding to public health systems in many countries, Gavi contributes to, among other things, the protection and training of health professionals and to the tracking and purchase of diagnostic tests. Additionally, as mentioned in the previous chapter, Gavi is leading COVAX, an international initiative that quickly identified the most promising anti-SARS-CoV-2 vaccines for the poorest countries, accelerating their development, production, and distribution, while also trying to guarantee equal access to the vaccines. Other COVAX partners include the CEPI, WHO, and the Bill & Melinda Gates Foundation.

Gavi's remarkable achievements are not cause for complacency. Too many children in the world lacked access to immunization through routine vaccinations, even before the COVID-19 emergency. The destabilization caused by SARS-CoV-2 requires an even greater effort to recover lost ground and resume with even more energy than before. New investments are needed to close the gap in childhood vaccinations between rich and poor countries.

One truly extraordinary result Gavi is aiming for is to eliminate those decades that separate the commercialization of a vaccine in a rich country compared to poorer countries. In other words, it is trying to ensure that countries of the Global South can benefit from the most advanced technologies and vaccines as they become available in the richer nations.

## Philanthropists committed to infectious diseases

We cannot mention Gavi without spending a few words on the Bill & Melinda Gates Foundation. Since its establishment in 2000, the Foundation has been committed to fighting infectious diseases that particularly affect the poorest populations around the world. Bill Gates and Melinda French Gates, at the time of their marriage, pledged to give half of the wealth they accumulated through Microsoft's business success to charity. They also tried to persuade other very wealthy people to follow their example.

Over the last twenty years, the Bill & Melinda Gates Foundation has spent some $53.8 billion, 30 percent of which is dedicated to global health. In addition to the activities just described with Gavi, the Bill & Melinda Gates Foundation have pledged to accelerate steps towards eradicating malaria, fighting AIDS and TB, and eliminating polio, areas in which they are making a tangible difference.

However, the Bill & Melinda Gates Foundation's innovative approach is not limited to its colossal financial commitment. Its funding is informed and driven by the most solid and advanced scientific evidence, by an effective merit-based process for allocating resources, and by the analysis and evaluation of results. Its vision has also changed methods and approaches used by many other international entities active in the global health sector.

Let us take just one example of how the Bill & Melinda Gates Foundation operates. The challenge of keeping vaccines at optimal temperature can be a problem even for rich countries, as we have seen with some of the mRNA-based COVID-19 vaccines that must be kept at very low temperatures. The same is true for conventional vaccines that need to be stored between 2 and 8°C, with delivery to some of the most remote places on Earth. Here the roads, if they exist, are impenetrable: electricity and refrigerators very rare commodities. If the vaccines get too hot or too cold they become ineffective. The optimal temperature must be maintained during all phases of the so-called "cold chain" (Figure 2). The Bill & Melinda Gates Foundation has launched calls for inventors to develop lightweight and inexpensive portable refrigerators that can

The long road to vaccination

Vaccines must be kept between 2–8°C all the way
from the factory to some of the most remote places
on the earth.

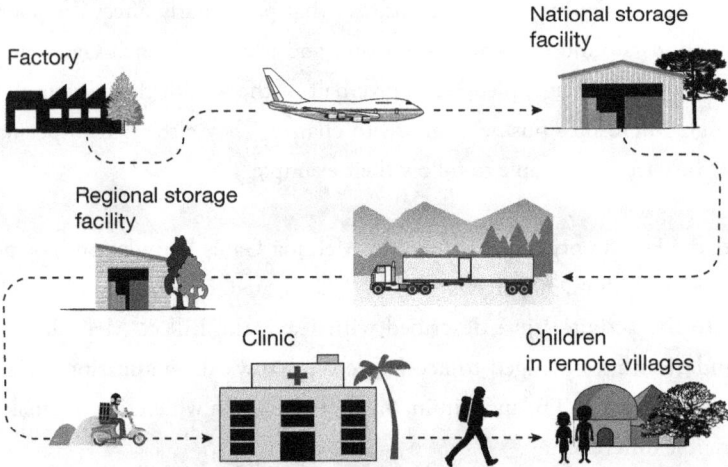

Figure 2: The cold chain for vaccines (adapted from *Gates Notes*).

maintain their temperature for a long time without the need for ice, batteries, or electricity. This initiative has resulted in several prototypes, some of which have gone into production and distribution in various countries, particularly in Africa, increasing the likelihood of vaccinating even for those children born in the remotest villages.

Bill Gates and Melinda French divorced in 2021. Hopefully, their personal issues will not change their philanthropic intentions, and the Foundation's activities in favor of vaccines and global health will go on with their now habitual contributions and energy.

## When the for-profit sector also does nonprofit

In the catalogue of contagious diseases for the Southern hemisphere, we briefly mentioned diarrhea. Yet this disease kills more children under five than malaria and AIDS combined. Two vaccines are currently being developed at the GSK

Vaccines Institute for Global Health (GVGH), founded in 2008 and based in Siena. One is a polyvalent vaccine against three forms of *Salmonella*, and the other is a bivalent against *Shigella* and a toxic form of *Escherichia coli*. These are some of the germs that cause the approximately 1.7 billion cases of diarrhea each year.

In 2010, the GVGH reached an important milestone, bringing a typhoid fever vaccine candidate to the clinical stage in just two years. The dual purpose of this vaccine is to fight a serious children's disease in developing countries and to stem the advance of an antibiotic-resistant bacterium all over the world. The technology developed was transferred to an Indian company that can produce the vaccine at an affordable cost, and this same company, in collaboration with GVGH, obtained the Indian government's approval in 2020 and the WHO's in 2021. The vaccine is now recommended for developing countries by the WHO.

Another of GVGH's targets is a vaccine against Group A *Streptococcus*, which is the cause of some serious and invasive infections and complications, especially in the most neglected regions of the world.

## The last mile

Resources for vaccines are never enough, but we can at least say that for the developing world, some very important assets had been secured even before the COVID-19 crisis. First of all, a lot of excellent science and technology has been made available; money and vaccines supplies have become more abundant than in the past; and agreements between international organizations and governments of developing countries have been signed. Still, we have yet to cover the so-called "last mile": a few million people yet to be immunized, mostly children, scattered around the most diverse places on the planet.

Even in its most literal sense, the last mile is never trivial. Dorando Pietri won the marathon in the 1908 London Olympics, only to be disqualified. The umpires had helped him up when, exhausted, he had more than once fallen in the last hundred meter stretch before the finish line.

The "last mile" can be fatal to vaccines too, even in rich countries, if all details are not carefully planned. We have witnessed the chaos and the delays that have characterized the initial phases of vaccination campaigns against COVID-19 in more than one nation due to poor organization. In underdeveloped countries, the problems that need to be considered are different, but no less complex.

The twenty-year experience of Les Cultures, an Italian non-governmental organization providing medical assistance in northern Niger, is a good example for appreciating what needs doing in practical terms to assist local doctors and nurses in vaccinating children.

First of all, you need to understand that most Nigerien mothers lack access to a convenient clinic a few minutes' drive away. The nearest health centers are usually tens of miles away, a distance that must be covered on foot, carrying their children on their shoulders. The long walk is generally along sandy desert tracks that can be erased by a single downpour.

It is all so difficult that mothers generally do not take their children for their shots, unless vaccinators come to their villages. This is what local health professionals do, with their "mobile vaccination unit" provided by Les Cultures. The lofty name hides a very basic off-road vehicle that carries vaccinators and their materials from village to village. The makeshift clinic is created on the spot, under the broad shade of a beautiful acacia tree, as in Figure 3 (obviously, vaccinations take place only when it is not raining).

In immunization clinics of wealthy countries, you frequently find brochures and other materials, explaining how useful and important routine shots are for children. This kind of communication does not work with Tuareg mothers. The Tuareg are a nomadic ethnic group that inhabits a vast area of the Sahara, including Niger, where they largely live from agriculture and herding. Only two or three out of ten Tuareg can read or write, men more so than women. Niger's official language is French, but most people speak dialects that are unique to the village they live in.

Figure 3: Vaccinations under an acacia tree in Northern Niger (Les Cultures).

The high illiteracy rate and the dialects are not the only obstacle to effective communication. Many people living in these villages have beliefs about individual fate being governed by magic and spells. How can vaccinators convince mothers that shots, and not witchcraft, are a true lifesaver for their children? How can they fix appointments for boosters when there are no clocks or calendars?

All these issues are far from simple. However, the health professionals working there have gained these mothers' trust, helping them understand that vaccinating their children is the best thing they can do to help them survive beyond the age of five. Today, one out of twelve Nigerien children does not make it to that age, and life expectancy is sixty-two years, twenty years short of the United Kingdom's.

Once the mothers are on board, vaccinators can begin giving the children their shots. The syringes provided by Les Cultures are self-destructing after use (another Gavi initiative), reducing the risk of HIV infection or other diseases from contaminated needles.

Everything is for the best, so to speak, until some unexpected event comes to jeopardize this delicate organization, whether it is guerrilla warfare between different ethnic groups, or one of many widespread criminal activities, or even a pandemic that arrives like a tsunami to disrupt the entire world. Persistent and patient, vaccinators will once again roll up their sleeves and begin from scratch.

## What challenges will vaccinology face after COVID-19?

More than ten years after the first edition of this book, published in Italy in 2009, we have updated this brief excursion into the world of vaccines. As we were writing, our homes, offices, country, and indeed the entire world were devastated by a pandemic. Our immune systems were not primed to recognize the new virus, and so our human society was left unprepared for the social and economic destruction inflicted by this disease of epidemic proportions.

Over roughly half a century, advances in the field of vaccines have followed one another with intense conceptual and technological revolutions. Every five years within the last twenty have seen complete revolutions in the technical tools available and the possibilities for action. It all began with raw preparations which contained whole germs, of which little was known, except how to kill or attenuate them. Yet even with those pioneering vaccines, billions of people were protected from all the diseases caused by stable germs, such as smallpox, polio, rubella, measles, and diphtheria, to name just a few.

The genomic revolution, which enabled reverse vaccinology, made it possible to design vaccines against microbes whose appearance changes often, such as meningococcus and the influenza viruses.

Today, antigens for vaccines are more precisely selected and optimized, thanks to the use of monoclonal antibodies, and to the progress made in the fields of structural vaccinology and immunology. It goes without saying that, behind all this progress are the many techniques that enable the easy introduction of changes in genes.

Finally, the techniques developed by synthetic biology and gene therapy have led us to the extraordinarily rapid development of mRNA vaccines and adenoviral vectors that promise to get us out of the current pandemic (Figure 4).

What challenges will vaccinology face, once the COVID-19 crisis is over? The global public health emergency has stimulated an acceleration of at least a decade on the technologies used to develop vaccines, and this is a very useful legacy for the future. We can hopefully use them to create new vaccines against antimicrobial resistance: a slow but inexorable pandemic in the making, according to many observers, unless decisive action is taken.

The same techniques can also be used against chronic infections. In these cases, cells of the immune system have been already *primed*, i.e., they have already encountered the antigens, but the body's defenses have been unable to completely eliminate the pathogen. As a compromise, the immune system has entered into a tenuous pact of nonbelligerence with the pathogen. This latent infection persists silently, without causing the disease, at least until other problems arrive to disrupt the equilibrium (this happens, for instance, with TB). Vaccinologists do not yet know how to immunize an organism against this type of chronic, ongoing infection. The recently developed vaccine against *Herpes zoster*, with a better choice of antigen and a good adjuvant, bodes well for other pathologies too.

Even more difficult is the challenge to develop therapeutic vaccines against those chronic diseases, where the immune system has, so far, always been defeated by the infectious agent (this is the case with AIDS). No vaccine has been successfully developed yet for diseases of this kind. Here, advances in cancer immunotherapy, where an immune system numbed by cancer has been successfully reawakened, can help with vaccine development as well.

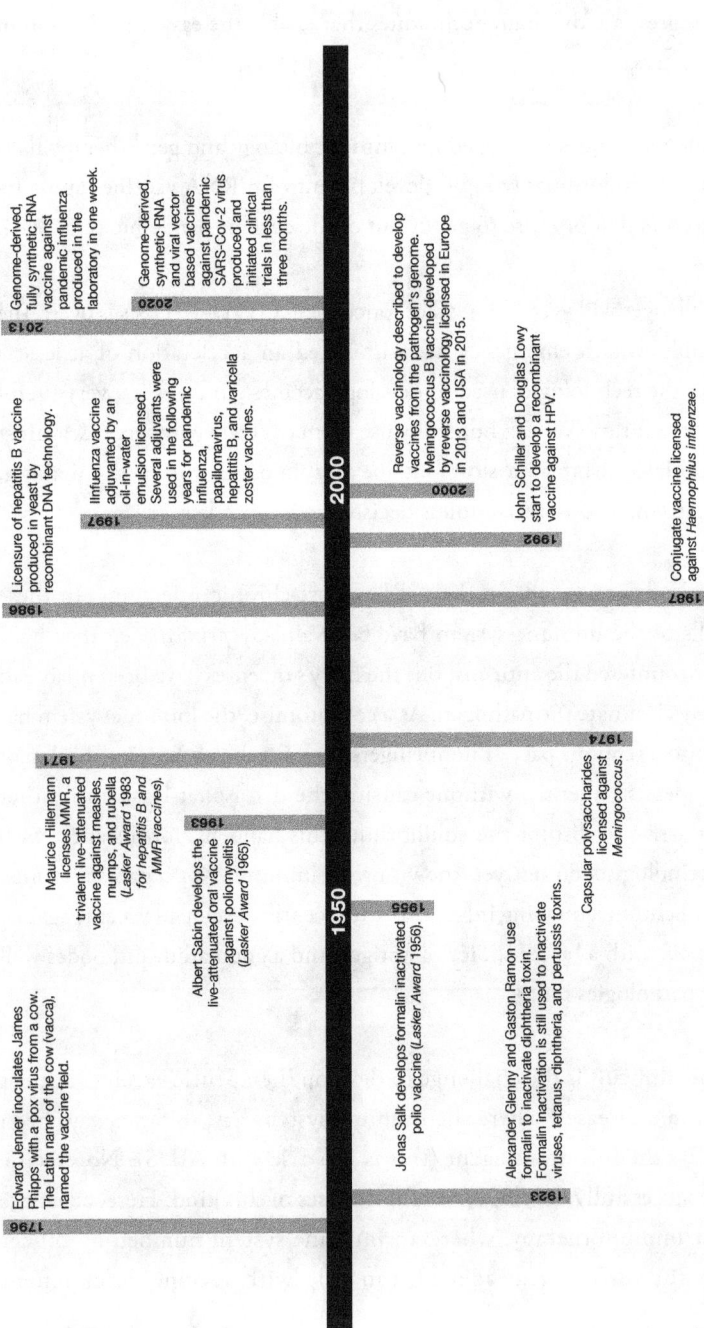

**1796** — Edward Jenner inoculates James Phipps with a pox virus from a cow. The Latin name of the cow (vacca), named the vaccine field.

**1923** — Alexander Glenny and Gaston Ramon use formalin to inactivate diphtheria toxin. Formalin inactivation is still used to inactivate viruses, tetanus, diphtheria, and pertussis toxins.

**1955** — Jonas Salk develops formalin inactivated polio vaccine (Lasker Award 1956).

**1963** — Albert Sabin develops the live-attenuated oral vaccine against poliomyelitis (Lasker Award 1965).

**1971** — Maurice Hillemann licenses MMR, a trivalent live-attenuated vaccine against measles, mumps, and rubella (Lasker Award 1983 for hepatitis B and MMR vaccines).

**1974** — Capsular polysaccharides licensed against Meningococcus.

**1986** — Licensure of hepatitis B vaccine produced in yeast by recombinant DNA technology.

**1987** — Conjugate vaccine licensed against Haemophilus influenzae.

**1992** — John Schiller and Douglas Lowy start to develop a recombinant vaccine against HPV.

**1997** — IInfluenza vaccine adjuvanted by an oil-in-water emulsion licensed. Several adjuvants were used in the following years for pandemic influenza, papillomavirus, hepatitis B, and varicella zoster vaccines.

**2000** — Reverse vaccinology described to develop vaccines from the pathogen's genome. Meningococcus B vaccine developed by reverse vaccinology licensed in Europe in 2013 and USA in 2015.

**2013** — Genome-derived, fully synthetic RNA vaccine against pandemic influenza produced in the laboratory in one week.

**2020** — Genome-derived, synthetic RNA and viral vector based vaccines against pandemic SARS-Cov-2 virus produced and initiated clinical trials in less than three months.

Figure 4: The timeline summarizes the main steps that led to the development of increasingly innovative vaccines capable of defeating infectious diseases (scheme adapted from *Cell*).

We could fill an entire encyclopedia listing the challenges for tomorrow's vaccinology, and we would not even be scratching the surface. And so, for this is where we stop, we hope that our readers have enjoyed this brief journey into the world of vaccines, having found in it some useful information along with a small taste of the pleasure a vaccinologist experiences when inventing new useful tools for human health.

# Appendix:
# How Do Our Defenses Work?

Our body's defense mechanisms protect us from disease and have a long evolutionary history. Whenever a germ plays cops and robbers with our organism, the game involves complexities, red herrings, and twists worthy of an Agatha Christie novel. But the "good guys" do not always win: if the attack is particularly ingenious the germs can prevail, but where the defense strategy is more brilliant, in the end, our organism will prevail.

Through the course of time, the weapons that have proven to be the most effective against invaders are preserved in the great arsenal that is our immune system. This arsenal includes a vast repertoire of tools developed during our evolution and is tested by natural selection every time our body launches a defensive action against an enemy.

What kind of weapons are we talking about? Objects forged with iron and fire, explosive powders, H-bombs? Not at all. The objects we are discussing are invisible to the naked eye but are still effective despite their size. They are the cells and molecules of the immune system.

These include conventional tools of protection, which are the same for everyone, as well as flexible devices, capable of adapting to the specific circumstances and encounters of individual lives.

But weapons alone are not enough. Rules, security procedures, communication systems, postal services, and intelligence operations allow the defense system to strike "enemies" so accurately and precisely that all armies on the planet pale in comparison. A virtually unlimited number of troops, and the ability to continuously mobilize new recruits, would be the envy of every general.

Unraveling the endless number of cells and molecules involved is a very difficult task, not to mention the very complex networks and interactions, and the huge variety of defensive actions available. Immunologists, who have spent the last two centuries or more trying to understand how the immune system works, know something about it.

The challenge is immense, equal at least to the difficulty of understanding how the brain works or of sending humans to Mars. But difficulties generally excite scientists, who are stubborn types. And the stakes are very high, with the Nobel Prize awaiting those who resolve the most perplexing conundrums. More importantly, life becomes an alternative to death for those who can avail themselves of new therapies and vaccines that broaden our arsenal for immunity.

The continuous acceleration experienced by experimental immunology has led to an accumulation of more discoveries over the last thirty years than in the previous two hundred. However, the knowledge we have today looks a bit like a map made of leopard skin, where the spots represent the best understood topics, for which we have a very large amount of detailed information. Between one spot and the other, we know that important links and connections are in place, even though we are not always able to explain exactly how they work. Finally, the background is a very vast area of things that we still do not know, but hope to understand in the future.

Over the next few pages, you will not find a catalogue of all the cells and molecules that operate in the immune system; there is just not enough space. You would also be quickly discouraged by the nasty habit immunologists have of assigning incomprehensible names to all they discover (often obscure, even to insiders). Finally, the catalogue would be obsolete even before going to print, because not even the fastest compiler can keep up with the frenetic pace of discoveries.

By leaving out most of the incomprehensible acronyms and nonessential details, we will try to give you an overall picture of the logic behind how our defenses operate. From this brief excursion into human immunity, you will see how this extraordinary and unparalleled defense system can function thanks to, in particular, its specificity, its effectiveness, its flexibility, and its redundancy.

## Physical barriers

Walls are a shrewd invention of the ancient world. Unlike their modern counterparts, that are mostly symbolic and erected by brutal, tyrannical, and reactionary regimes, walls in the fortified cities of ancient times allowed human societies to develop trade and the arts safe from unwelcome visitors.

Living beings too have erected "walls" in their own way in the course of evolution. These biological walls are the skin, the mucous linings, the cell membranes, and the vesicle linings found inside the cells. Altogether, they create many distinct and protected environments, optimal conditions for the development of complex functions.

If we imagine our body as a region similar to those studied in geography, the most obvious border between us and the outside world is made up of skin, a large surface that in adults is roughly equivalent to two square meters and that covers every organism.

But we also have other walls in contact with the external environment in addition to skin. These are the mucous membranes covering the deep inlets and branched fjords that penetrate the digestive tract, the respiratory system, and the reproductive system, for a total area of about four hundred square meters. If we were to spread the skin and mucous membranes out on a surface and compare their breadth, we would have a medium-sized blanket (the skin) alongside a tennis court (the mucous lining).

Beyond these barriers, food and other precious resources thrive, making the human body a seductive terrain of conquest for armies of bacteria, viruses, and parasites. Entering is not easy, the defenses are resistant. Sometimes though, an invader succeeds in breaking through, for example, via a wound, and getting inside.

## Defenses beyond the walls

Having broken through the physical barriers, the invader must contend with a barrage of various biological weapons, the most effective of which is most probably the *complement* system.

Present in massive doses in all body fluids, the complement system consists of a few dozen proteins that act by means of a cascade effect to destroy invaders and eliminate cellular debris. In practice, the action of these proteins could be compared to a flamethrower swiftly attacking anything it finds in its way.

Human cells are saved from this destructive action thanks to a special protein "shirt" with which they are all uniformly coated. Just like a football jersey, this coating allows the complement to recognize the host's own tissues.

In this way, the complement can accurately distinguish between what it needs to attack (the invaders, without the distinctive human "team jersey") and what it is wiser to save (our cells, covered with special protection). This protein-based shirt is even supplied by the complement, almost as if it were aware of its capacity to inflict significant damage.

As additional protection against attacks from complements, human cells expose some molecules on the surface which, like bow spurs on ancient warships, can neutralize any attack.

It is likely that the complement is the oldest defense tool available to the human species. In fact, something similar is at work in simple multicellular organisms, like corals and sponges, which have at least one billion years of evolution behind them.

Macrophages constitute another front line weapon: an army of cells present in all solid tissues, located just below the skin and mucous membranes (Figure 1). Macrophages are stationed like sentinels guarding the outposts of the human body, intent on intercepting any invaders.

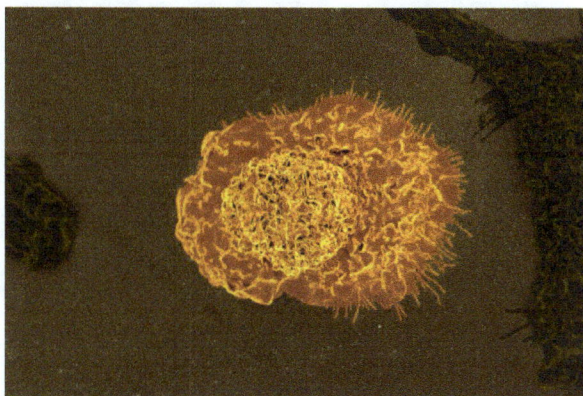

Figure 1: A macrophage (Wikipedia).

Macrophages are equipped with special "antennae," called *receptors*, positioned on their outer membrane, that can "sense" certain distinctive signals associated with most routine invaders. For example, one of these signals is mannose, a sugar used by many microorganisms, but not by the human body, and which the immune system's cells recognize as a sign of danger.

When they sense the presence of a bacterium, the receptors push the macrophage to move, and it then slithers towards the enemy, thanks to extensions called *pseudopodia*. On reaching the bacterium, the macrophage enfolds it, forming a kind of pocket called a *vesicle*. The vesicle is then ingested by the macrophage.

The entire process is called *phagocytosis* and is a primordial feeding strategy, still used today in its original incarnation, for instance, by amoebas. These unicellular animals have been capturing their food in this way for a few billion years.

The work of macrophages is not limited to eating debris or bacteria. Macrophages are also able to receive danger signals from other components of the immune system and, in turn, ask for reinforcements needed to address infections and other critical situations.

Sometimes the signals received by the macrophages are so alarming, that they start producing molecules of reactive oxygen and phagocyte substances in greater quantities than usual.

With such powerful weapons, even the most aggressive bacteria generally do not stand a chance against these "big eaters." But what happens if an invader is too large to be swallowed up? Well, the macrophages will release powerful poisons onto it.

The immune system is also equipped with on-call assassins. These cells, known as *neutrophils* and *natural killer* (NK) cells, circulate in the blood, waiting on a summons from components of the system to disarm and kill dangerous microorganisms.

## Conventional weapons we are born with

Macrophages, neutrophils, NK cells, and the complement system are effective members of the oldest evolutionary line of defense, one we share with most animals. This is a legacy, common to all, which accompanies us from birth, and for this reason is called our *innate immune system.*

They are all versatile and effective tools for destroying invaders swiftly while protecting the host's tissues, eliminating cell debris, and calling for reinforcements when needed.

The innate immune system defends us reasonably well from bacteria and other germs that our species has known for millennia, and that we can expect to encounter in the course of an individual life. It is an effective, but nonspecific, generic system: a germ is not recognized individually, but through signals—for example, mannose or the lack of complement coating—common to numerous other invaders.

Taken together, all the components of the innate immune system share a common goal of destroying the enemy in the shortest possible time, because without their intervention, once a bacterium has penetrated inside our body, it can reproduce up to a few trillion units in twenty-four hours.

When the innate immune system is successful, victory is so effortless that we barely notice it. But there are times when it does not work, perhaps because the invader has learnt to pass itself off as the host, for example, by mimicking the complement's protective coating that is specific to human tissue. Almost all the attackers that have forever haunted our innate immune system attempt some form of mimetic strategy to evade surveillance. We could say that the innate immune system's only real limitation is the venerable age of its strategies: today our first line of defense tactics, while still very effective, are often circumvented by most attackers, who manage to slip through.

What happens when the innate immune system is unable to contain an aggression? It launches the equivalent of an SOS and calls for reinforcements, through the immune system's mail and communication networks.

## Molecular communication

Attacks by invading germs can be simultaneous and multiple, and spread throughout the body. Luckily, the human body possesses a highly effective communication system that coordinates defenses against multiple attacks. Let us see how it works.

The immune system's cells exchange messages by using special molecules called *cytokines*. Cytokines carry quick, clear, and unambiguous instructions. They signal alarms, but they also signal that the coast is clear, and when reinforcements are needed they ask the immune system's cells to reproduce at high speed, or they shout "stop proliferation!" when the danger has passed. They issue and remove the license to kill.

Let us take a look at a concrete example. A dangerous invader has penetrated inside our organism and is plainly reproducing itself. The complement system, the macrophages, and other cells have intervened promptly, but their action is insufficient: the germs continue to reproduce at a rapid pace. That is when all the cells involved release cytokines to send out an SOS: "We are looking for strong killers to disarm and terminate dangerous subjects."

Cytokines circulate in all body fluids, reaching other components of the immune system that can help defeat the infection. These other components are mostly immune cells that respond promptly and begin travelling in the blood. But how do they know where to go?

## Circulation and postal service

For centuries, human societies have resorted to waterways as the main communication channel for goods and people. The interior of our body, like that of other mammals, works in the much same way.

The bloodstream, with its capillary structure, is in fact the main navigation network for reaching every region of the body, even the most remote. Our body is made up of something like a hundred thousand billion cells, and each one of these cells is never more than half a millimeter away from a blood vessel. At any greater distance, a cell would be unable to access oxygen and nutrients, or eliminate waste, and would therefore die.

The immune system exploits the blood flow to deliver defense tools and instructions. They are not unlike billions of "parcels," constantly circulating in all different areas of the body. Since the circulatory network is very extensive and navigation is quite fast—at a blood rhythm of sixty complete circuits per hour—how does each parcel know at which station to alight? The immune system has solved this problem through a special postal service based on adhesive particles and molecular codes.

To better understand how this system works, let us again look at a concrete example. Macrophages engaged in fighting an infection call for reinforcements by launching some cytokines. Neutrophils and other cells sense the call and post an adhesive molecule on their membrane, which can be identified by a kind of code, also made of molecules.

As they travel through the circulatory system on their 'mission', these cells run into other adhesive molecules—carrying the same code—that are attached to the blood vessel wall. The related molecules recognize each other and hook up. The cell is now stuck to the blood vessel and is forced to slow its pace, rolling along the vessel wall. In practical terms, this slowdown stops the cell at the point where it needs to leave the blood vessel and penetrate into the infected tissue.

The adhesive molecules and the molecular codes are a bit like the addresses we write in email messages, and they are just as, or even more, efficient. As long as the organism is sufficiently young and agile, the immune system's cells can reach the right "exit" to access the site of infection without a hiccup.

## Generators of molecular diversity

The innate immune system works so well that most of the time we do not even notice that we are under attack, as long as the invader is not skillfully disguised, pretending to be something that belongs to us. When this does happen, our immune system is equipped with an immense and very diverse repertoire of special proteins, called *antibodies*, which can adapt to every possible substance in the universe, as well as function as molecular "antennae."

This potentially infinite collection of antibodies is produced by *B lymphocytes*, real factories of molecular diversity at the service of our immune response. You may wonder why these cells are named this way: the letter B is the initial of *bursa of Fabricius*, the bird organ in which they were first identified. Following the discovery of B lymphocytes in humans as well, the letter B has remained to indicate that in our body they are produced in the *bone marrow*. It is precisely there, in the soft tissue located in the cavity of our long bones, that B lymphocytes are born and grown.

Each one of these cells is a specialized factory, assembling only one type of antibody that specifically recognizes a single foreign substance, called an *antigen*.

We have seen that antibodies are proteins, and that each protein is built according to a distinct instruction stored in the genes within the cell nucleus. Given the large number of different antibodies we have, you would expect the instructions dedicated to this vast and diverse production to take up a large part of the genetic material, but this is not the case. These instructions occupy a very small part of our genes, too small to account for all that variety. There must be a trick, but what is it?

Take the Lego® bricks: with just a limited variety of pieces, a child can build a vast number of different structures. The B lymphocyte production system works in much the same way: a varied interlocking of pieces of protein antennae, from a limited number of genetic instructions, makes it possible to produce a vast array of molecular receivers.

The genetic mechanism responsible for this variable interlocking production system is actually a little more complex than what we have just described. Rather than delving into the biological details of it all, let us focus on the properties that this mechanism provides the immune system with: maximum adaptability, flexibility, and specificity in the face of enemies, with specific weapons for each possible invader, available with an extraordinary economy of resources. A diverse and limitless repertoire of antibodies built from a minute amount of genetic information.

## Defenses on demand

About three billion B lymphocytes circulate in the blood, and of these, about one hundred million each produce a different kind of antibody. We now have a small math problem to solve: how many individual B cells are in our arsenal against every invader?

Dividing the total number of B lymphocytes (three billion) by the number of different kind of antibodies (one hundred million), the result we get is thirty. Thirty B lymphocytes per antibody type is the minimum dose from which, during an attack, the immune system can start from to rapidly produce an adequate number of identical cells for use against the enemy.

How does this happen in practice? The B lymphocytes circulating in the blood have many identical antibodies exposed on their membrane. Each antibody has a characteristic Y-shape. The base of the Y serves to anchor the antibody to the cell membrane, while the two upper ends of the Y are the parts that recognize the antigen, and attach to it.

Like those in search of a soul mate, every B lymphocyte spends most of its lifetime looking for its own antigen. The probability of encountering the antigen is very low. But sometimes a meeting does take place and the two proteins are joined. The bond formed is very specific and secure, because both ends of the Y each have a complementary shape with respect to the antigen (Figure 2). It is so specific that the close relationship binding an antigen and an antibody is often compared to the union between a key and its own lock.

The bond with the antigen induces the B lymphocyte to double its volume by splitting into two daughter cells. These divisions are continued by daughter cells, daughters of daughters, and so on, for about a week. Proliferation proceeds at the rate of a cell division every twelve hours: when it stops, more than fifteen thousand new B lymphocytes have been produced, all identical

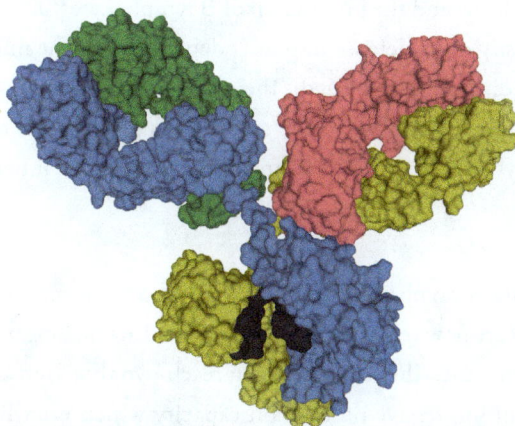

Figure 2: Scheme of an antibody molecule. The typical Y-shape allows the antibody to anchor to the cell membrane, with the lower end, and to recognize the antigen and bind to it, with the two upper ends (Wikipedia).

to the original cell. This set of cells, which descend from the same progenitor cell and produce antibodies of the same type, is called a *clone*. An immune system whose lymphocytes have not yet encountered a certain antigen is said to be *naïve*, while one that has encountered it is said to be *primed*, or ready to react.

The B lymphocytes strategy of "proliferation on demand" is one of the smartest that the immune system has evolved. It allows for the production of a large quantity of specialized cells against any given invader, once present, without the need to stockpile in the long periods of "peace." The saving is primarily one of resources (we do not keep cells alive when they are not needed!), but also of space (the broadest possible variety of B lymphocytes, ready to proliferate in the event of an attack, can be stored in the blood).

In any self-respecting army, the number of troops that can be mobilized must be equal to those deployed by the enemy: it cannot be too low, or defeat is assured, but it cannot be too high either, otherwise the reserves are depleted. The immune system can also accurately calculate just how many resources it needs to produce. The greater the number of antigens present, the more intense the proliferation of specific B lymphocytes. The ratio between the number of antigens and the production of B lymphocytes allows the body to vary the intensity of its defense response, depending on the number of dangerous particles to be neutralized. The criterion of proportionality of the defense, as against the degree of the offense, is a general rule often used by the immune system to regulate the strength and resources it needs to deploy in its actions.

Once the proliferation phase is over, each B lymphocyte begins to manufacture antibodies that are now released into the blood, and are no longer kept attached to the cell membrane. The release rate can reach two thousand antibodies per second. It is an impressive production capacity which each B lymphocyte maintains for about a week, before dying from exhaustion. But no need to worry, new blood cells are continuously being produced to replace those dead from exhaustion or eliminated.

## The role of antibodies

Once in the blood, what do the antibodies released by B lymphocytes do during an infection? They attach themselves like suckers to the antigens they find on the enemy.

For an invader to find itself with so many antibodies attached to its own antigens—we are talking thousands—is the equivalent of being tagged on a photo on the most viewed page of Instagram. Except that, instead of a 'like,' the antibody label carries with it a death sentence for the invader.

The death sentence must be carried out, but by whom? As alert and sticky as they are, antibodies are rather harmless folk. Their role is to attract the attention of the complement system, macrophages, and other cells more suited to causing the death of the invader.

So far, we have described antibodies as soldiers of the immune system, typically Y-shaped and capable of attaching to antigens, as well as being indistinguishable from each other, save for their ability to singly bind to a different antigen. Reality is always a little more complex; in the case of antibodies, nature has invented five useful variations. *Immunoglobulins M* (IgM), formed by five Ys joined in a large star-shaped molecule, are the oldest antibodies from an evolutionary point of view (fish have them too), and the first that B lymphocytes release into the blood. IgM is also particularly effective in activating the components of the innate immune system for destroying the invader.

Following closely after the IgM, we get the *immunoglobulins G* (IgG): formed by a single Y; they make up about 75 percent of the antibodies circulating in the blood and live longer than all the other classes of immunoglobulins (three weeks, versus a single day for IgM).

*Immunoglobulins A* (IgA), formed by two Ys attached by their "feet," guard the mucous walls of the body, including the intestine, where they intercept invaders while resisting waves of acids that flood our digestive tract; IgA are also the protective antibodies in newborns, obtained from maternal breast milk.

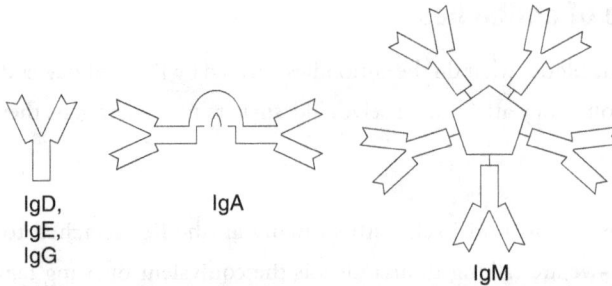

Figure 3: Schematic representation of the five classes of antibodies.

*Immunoglobulins E* (IgE) protect us from parasites and can trigger allergic reactions. Finally, *immunoglobulins D* (IgD), made of a single Y unit, are the fifth and last class of antibodies, and their function is still unknown.

Figure 3 summarizes the five classes of antibodies; we have all read at least one of these abbreviations in blood test results (the concentration of the different types of immunoglobulins in the blood under normal conditions is within defined values; variations from these values indicate that something is wrong).

Thanks to the specific action of antibodies, our body can label each germ so that other agents within our immune system will deal with it and its progeny in a definitive way. And we even know how to paralyze certain viruses which, overrun by antibodies, are blocked from penetrating our cells. But should a virus escape this control and enter a cell, our antibodies are no longer of any help. The cell membrane is like an opaque wall, hiding everything that happens inside. So, we will need other weapons against these infections, that can see beyond the "wall."

## Viral pieces on display

Take a thief who decides to rob a country villa while its owners are away on holiday, with both the caretaker and guard dogs soundly asleep and the alarm system deactivated. The thief can freely move around because no one sees him, no one hears him, and no signal warns the police of his presence.

A virus that has just entered a host cell is a bit like that thief; the difference being that, rather than wanting to steal the cell's resources, it is more interested in using them to produce thousands of copies of himself.

How does a virus replicate? To understand this, we first need to get an idea of how it is made. A viral particle contains its own genetic material—DNA or RNA—wrapped in a protective envelope made up of proteins. The genetic material contains the instructions needed for producing new viruses, while the protein envelope is essential for both protection and to penetrate the cell during infection.

Compared to a cell, a virus is a simpler structure, lacking several things: not least the machinery needed to build copies of proteins and genetic material and assemble them into new viral particles. In order to do this, the virus "hijacks" the host cell's machinery, forcing it to almost exclusively manufacture genetic material and proteins destined for viral progeny based on the instructions contained in its DNA or RNA.

According to an old Italian proverb, "not all doughnuts come with a hole," which adapts well when describing the material produced for making new viruses. For each optimally assembled particle, several imperfect molecules are discarded by the cell factory's quality control system. What happens to these scraps? Most of them end up being broken down and reused by the cell. But for the many residual protein fragments, known as *peptides*, there is also another possible scenario.

The *major histocompatibility complex* is a special protein, whose complicated name can fortunately be abbreviated to MHC. Its function can be understood with a political metaphor. In the days before electoral campaigns were conducted almost entirely through newspapers, television, and social networks, trucks bearing giant posters of candidates' names and faces would be driven around the city. The MHC protein does something similar with the discarded viral peptides: it finds them inside the cell, and if it identifies them as sufficiently different from the organism's usual molecules, it picks them up and carries them to the cell's external surface. Here, the MHC with one "hand" exposes the viral peptide while remaining anchored to the cell membrane with its "foot."

But what is the point of exposing the viral peptide on the surface of an infected cell? So that it can be spotted by one of the immune system's "special investigators." The sighting will alert the body that an infection is underway in the cell displaying the viral peptide, caused by the same virus that manufactured the peptide. If, on the other hand, a cell only displays its own antigens, it will not be attacked.

A cell that exposes a peptide on its outer membrane along with the MHC protein is, in immunologist language, an *antigen presenting cell*. In principle, all human cells can present an antigen, because each contains at least one form of the MHC and can therefore bring peptides to the surface. However, some cells are better at this than others: macrophages, dendritic cells, and B lymphocytes are absolute champions of antigen presentation.

## Virus hunters

Something suspicious is lurking inside a macrophage. Tangible signs of a viral presence emerge on the surface of the infected cell. These are viral peptides, transported by MHC proteins. Which cells can pick up on all of these clues?

*T lymphocytes* have a lot in common with B lymphocytes. Firstly, they produce a protein, the T cell receptor, that acts like a molecular antenna. For brevity, we will call it a TCR.

In many ways TCRs resemble antibodies: each T lymphocyte produces only one TCR type, specific to a given antigen, and a very large variety of TCRs are obtained thanks to a genetic mechanism analogous to that used by B cells to construct antibodies. Unlike antibodies though, TCRs are never released into the blood, operating while remaining attached to the T cell membrane.

We have about one trillion T lymphocytes in our blood, and ten thousand T cells for each type of TCR. Although each antigen has a very low probability of running into its own TCR, this does sometimes happen.

To increase the chances of an encounter, our body has a *lymphatic system*, which operates as a secondary communication network alongside blood circulation. The vessels of the lymphatic system collect the lymph, the fluid that passes from blood vessels to the tissues, and push it back into the blood under pressure exerted by the muscles. Lymph nodes, small spheres of dense tissue in which B and T lymphocytes and antigen-presenting cells are concentrated, can be found along the lymphatic vessels.

We have thousands of lymph nodes scattered throughout our body, making it more likely for an antigen to be intercepted by its own antibody or TCR, given the higher lymph nodes' concentration of B and T lymphocytes. The meeting between a TCR and its own antigen induces the T lymphocyte to proliferate. In about one week this gives rise to a populous clone of identical T cells, specific for that same antigen.

Given all these similarities, why do we need T cells? Are the B "cousins" not good enough? Living things are full of organs and structures that can do the same or very similar things, and the immune system is a perfect example of this. This *redundance* of interchangeable means of defense is not a case of waste by mother nature, who is generally rather frugal. It is, in fact, essential for dealing with invaders whose tricks are so many and varied that sooner or later, their ability to bypass the body's arsenal of effective weapons is pretty much inevitable. In other words, a good dose of redundancy serves to ensure that, if one sentry is sidestepped, others are still in place.

We have seen that antibodies are blunted weapons once a virus has infiltrated a cell. TCRs, on the other hand, can "see" a virus beyond the cell surface wall, because it is able to identify the traces exposed on the surface by the MHC. TCRs are particularly good at this because it recognizes antigens only if they are made of proteins, and only if they are attached to the MHC.

We all have three types of T cells. At the forefront are the *cytotoxic* T cells, which induce an infected cell to commit suicide once they have recognized a viral antigen presented by the MHC on its membrane. Unlike B lymphocytes, T lymphocytes are far from harmless.

As the rear guard, *T helper* cells coordinate the work of cytotoxic T lymphocytes and other agents, by sending out cytokines (Remember them? Cytokines are not just the immune system's messengers, they also help B lymphocytes proliferate and mature). As head of defense, the T helper cells are a lot more than just helpers, despite their name.

Finally, we have *regulatory* T cells, also called T-reg, whose function is not yet well understood.

Short terminology break: the letter T is the initial of *thymus*, a tiny gland located just below the breastbone, where T lymphocytes mature (B lymphocytes, as you may remember, develop instead in the bone marrow).

## Immunological memories

Juliet and Ahmed are siblings. When they were children, they got sick, at least twice, with the same disease. The first time was with flu: Juliet got away with a few lines of fever and one day in bed, while Ahmed took a week to get better.

The opposite happened with scarlet fever: Ahmed had a small rash and only one day of fever, while Juliet was forced to stay home from school for more than ten days.

Sometime after, Ahmed and Juliet spent a summer holiday in the mountains. Looking back to that holiday twenty years later, both remember it as one of the most beautiful of their lives, but for different reasons: Ahmed for the tennis court that began his long association with that sport, Juliet for an unforgettable romance.

In the course of our lives, the germs we encounter leave footprints not unlike the experiences we store in our memories. In both cases, their traces are accumulated over time and differ from person to person, even when the experience that generated them is apparently the same. Both the intensity and the duration of the experiences will combine to form memories.

Why do we have an immunological memory, and how is it formed? We have seen that the longest-living antibodies have roughly a three-week lifespan, which is more or less the same for B and T lymphocytes: from when they meet an antigen, to their death after they have given their all against the enemy.

It is a good thing that our body knows how to rid itself of obsolete weapons once the danger has passed. Think about it, if we were to hang onto every old cell or used antibody, we would end up exploding, poisoned by cells and molecules that are well past their due date. But removing all traces of the battles we have fought would be shortsighted, so our immune system hangs onto a tiny sample of B and T lymphocytes from each attack, which can be mobilized should the same enemy present itself in future.

The first time round, it takes about a week or two for the B and T lymphocytes to proliferate and produce antibodies and TCRs, more or less the average duration of an infectious disease. The second time around though, our defenses are activated so quickly that we usually do not even notice being under attack. Not only is the entire process faster, it is also more effective. In fact, after a first encounter with an antigen, B and T lymphocytes improve their performance thanks to a few genetic mutations in the cells' DNA, which increase the affinity between antibody and antigen, and between a TCR and an antigen.

The continuous creation of *immune memory* allows us not only to respond extremely quickly and effectively to each subsequent attack, but also to update the system's memory, just like a computer hard disk.

The ability to update knowledge about the attackers is what differentiates most the adaptive immune system, built on each individual's personal experience, from the innate immune system, which is the same for everyone.

Immune memory is at the basis of the effectiveness of vaccinations. Most modern vaccines contain an antigen, which is a small and harmless portion of a germ, or the genetic instructions for our cells to build such an antigen. When the vaccine is injected into the body, the B and T lymphocytes that meet the antigen will first recognize it, then proliferate, and finally die. However, a few of those cells remain alive, forming the immunological memory that is so useful in the event of a subsequent, real infection by the same invader.

## Safety systems and inhibitory brakes

By now you should have a rough idea of how the immune system works, but there is still one important piece to the puzzle that is missing. We need to say a few words about the entire system's safety mechanisms.

Why does the immune system need safety mechanisms? Is not an apparatus that coats everything that belongs to it in special molecules to protect them from getting caught up in the attack against germs safe enough? Can we not sleep soundly knowing that there is a system in place that guarantees a specific recognition between an antibody and an invader's antigen, or between a TCR and a fragment of viral protein?

The fact that the system's antigen recognition mechanism is highly specific is indeed proof that our immune system does not strike at random.

But let us assume that even just one B or T lymphocyte of the trillions circulating in our body gets it wrong. Within a week, over fifteen thousand new cells, identical to the misguided cell, could be ready to unleash a violent attack on something that is, in fact, totally harmless.

Fortunately, this scenario is highly improbable. Prudently, both B and T lymphocytes require one more activation signal, other than the one triggered by a bond with an antigen, before they begin producing and calling in reinforcements. This second signal, which is less specific than the first, involves the binding of two molecules distinct from the antigen/antibody bond, in the case of B lymphocytes, or from antigen/TCR bonding, in the case of T lymphocytes. It is a bit like having to insert a pin number once you have typed in your password when logging into your online back account. It is an additional check for ensuring the system's overall safety before triggering a reaction which, in addition to wasting precious resources, could produce inflammation, with undesirable consequences.

Another mistake that the immune system must avoid in every possible way is "friendly fire." To understand what this is, let us take an example: a viral antigen is exposed by the MHC on a cell's outer membrane. How does the TCR know that this peptide comes from the virus that infected the cell, and not from a protein belonging to our body? After all, all proteins are built in a very similar way, from building blocks—amino acids—which are the same in all living things.

It is a mistake that can theoretically happen, and when it does, the consequences are very serious: an autoimmune reaction. Fortunately, blunders of this type almost never occur because the B and T lymphocytes attend the *immune tolerance* school as they mature in the bone marrow and the thymus respectively. At this school, the only cells that pass their final exams are those that *cannot* recognize or attack the molecules belonging to our body. Only these cells can access the blood circulation and meet the antigens, while all cells that fail the test are eliminated.

In this way, immune tolerance prevents B and T lymphocytes from attacking the very body they are meant to defend. Rare errors in tolerance education give rise to so-called autoimmune diseases. In these cases, some antibodies mistakenly recognize molecules that belong to us and trigger aggressive reactions against them, causing serious damage and chronic inflammation in the tissues where they are located. Type 1 diabetes mellitus, or juvenile diabetes, is an example of an autoimmune disease where antibodies and lymphocytes attack and destroy the cells of the pancreas that produce insulin.

An additional safety check consists of inhibitory "brakes" pulled by the immune system once it has triggered a defensive action. What are these brakes for? They are very useful for preventing an attack from continuing indefinitely, well after the enemy has been defeated. An ongoing battle, if never stopped, risks doing dangerous and unnecessary damage to the body itself, for example by causing a state of chronic inflammation that can favor the onset of tumors and other diseases.

Today, we know how to use these inhibitory brakes to our advantage: James Allison and Tasuku Honjo deserved the Nobel Prize in Physiology or Medicine, in 2018, precisely for understanding how to remove the "brakes" on the immune system when "pulled" by a tumor, as it tries to protect itself from an attack.

Allison and Honjo's discovery, which led to the development of "checkpoint inhibitors," has revolutionized cancer therapy. A new clinical and research discipline, immuno-oncology, was born, and so far it has made several types of cancer more treatable, including melanoma and lung cancer. With these types of treatments, immunologists as well as oncologists assist in treating cancer patients.

But just like with cars, completely removing the brakes can be dangerous, and some cancer patients treated with checkpoint inhibitors develop autoimmune diseases and other issues as side effects from therapy. Still, this should not discourage either patients nor physicians as the search for remedies is ongoing. The history of medicine is paved with problems resulting from trying out new treatments, that have been gradually solved. The solutions come from the continued study and a better understanding of how our defenses work, and how we can exploit them to our advantage, to avoid harming ourselves.

## Let us summarize

The first line of defense an invader encounters are physical barriers, such as the skin and mucous membrane. Beyond these walls, that are difficult to penetrate, the innate immune system deploys its arsenal of weapons: complement, macrophages, NK cells, and neutrophils. They are present when we are born and are the same for all individuals. Some of these defense tools have accompanied life on Earth for over 500 million years and can also be found in very ancient species such as jellyfish. Their protection is very effective, but there are also numerous attackers who, with a millennial experience of these tactics, have learnt how to bypass this initial line of defense.

When the innate immune system alone is not enough, the soldiers of the adaptive immune system are called in: B and T lymphocytes, antibodies, and TCRs. These are refined tools, each specialized in recognizing a single and specific antigen. They are built with ingenious systems that allow for an immense diversity obtained from limited resources. They provide us with the ability to kill not only invaders lurking outside the cells, but also viruses that find refuge within the cell membrane. Identifying these viruses occurs when the MHC kicks in, by transporting and exposing pieces of viral proteins on the surface of an infected cell.

Unlike the innate immune system, the adaptive immune system continually updates its repertoire of knowledge about invaders, building a lasting immune memory that reflects each individual's encounters and experiences.

Host cells and tissues are protected from friendly fire because they are coated with complement proteins, and because the lymphocytes go to a school of tolerance before they begin working. Also, the security of the system is guaranteed thanks to multiple controls, like "brakes" and "cease-fire" mechanisms that are activated as soon as the infection is under control, as further attacks would be potentially dangerous and unnecessary.

The two systems, innate and adaptive, do not work separately, but collaborate intensely thanks to communications based on cytokines, shipping systems that use adhesion molecules and molecular codes, and meeting places such as lymph nodes.

Simplifying a bit, we can say that the effectiveness of the immune system depends on at least three factors: the *precision* of the cells and molecules involved in identifying and targeting an invading agent; the *number* of these cells and molecules that can be rapidly produced and deployed; and the *duration and intensity* of the action.

Most of the time the different elements of the immune system work in extraordinary harmony, like an orchestra where the strings play well together with the woodwinds, brass, and all the other instruments. We do not even realize it is happening. We only notice that something is not right on those very rare occasions when some mishaps happen: the immune system misses a target or hits out too little and too late (this sometimes happens in elderly people). Or it might act with excessive force (especially in young people). Or it might confuse something that is ours with a pathogenic organism and inflict self-harm.

The immune system is a very powerful mechanism, to be handled with kid gloves. Its very effective weapons, which we have briefly and incompletely described, are very useful if they are unleashed against pathogenic organisms and tumors. However, if they are directed against us, the outcome can be an autoimmune disease, brain inflammation, or a "cytokine storm" like the one that is killing so many patients with COVID-19.

As the British science journalist Ed Yong wrote, "the system may be vexingly complex, but it is also both efficient and resilient in a way that our society could take lessons from. It prepares in advance, and learns from its past. It has many redundancies in case any one defense fails. It acts fast, but has checks and balances to prevent overreactions. And, in the main, it just works. Despite the multitude of infectious threats that constantly surround us, most people spend most of the time not being sick."

Over time, we humans have come to understand at least some parts of this extraordinarily complex system. We have tried to explain and interpret the actions of its multiple cells, molecules, and interactions. And we have done our best to increasingly extend our natural immunity with vaccines, monoclonal antibodies, and immunotherapies. We have learnt to create and use vaccines, adopting them well before we understood the causes of infectious diseases and our bodies' defense mechanisms. The practice has developed and expanded to the point that it has eradicated many diseases that had previously been both widespread and deadly from our planet.

# Dispelling 7 Myths

**1. Vaccines can be useful, but they are dangerous.**
The number of cases of infectious diseases has dropped by more than 90 percent in just over half a century because they are prevented by vaccinations. Vaccinations were already safe and effective in the twentieth century, and they are even more so now, thanks to twenty-first century vaccine preparation technologies. In those rare cases where vaccines caused serious and untreatable side effects, they have been discontinued.

**2. Some vaccines can cause autism.**
The overwhelming majority of doctors and scientists agree that vaccines are not responsible for the autism that afflicts so many children. They can say this on the basis of studies of tens of thousands of people, duplicated in many countries.

Thiomersal, the mercury-based preservative blamed for causing autistic syndromes, was removed in 2002 from vaccines as a precautionary measure. There has been no sign of decreases in cases of autism since then.

**3. Childhood diseases are mild and not very dangerous: getting vaccinated is pointless.**
Many infectious diseases can cause serious problems in children, with life consequences. For instance, prior to a mumps vaccine, two out of one thousand children would end up in hospital, with roughly fifty children dying each year in the United States because of mumps.

**4. You can safely delay vaccinating your child until preschool.**
Several vaccine-preventable diseases are more dangerous when children are very young. Children can be exposed to infections even if they are not in school,

through contact with parents or siblings, while travelling, or even during a brief stop at a shop. They can also be exposed to diseases that are now rare, through contact with people who come from countries where vaccinations are less widespread. Protection during the first years of life is all the more necessary for these reasons.

### 5. If multiple vaccinations are administered with a single injection, they can damage the immune system.

It has been proven that a child's immune system can handle more than one vaccine at a time. Combinations of several vaccines cause immune responses comparable to those induced by the same vaccines when administered singly.

A child encounters thousands of antigens in a single infection, and during the initial five years of life a child will get sick between four and six times a year on average. All this far exceeds the set of antigens present in a full childhood immunization schedule.

### 6. Many vaccines are no longer needed because the diseases they protect against have virtually disappeared.

Many diseases have almost disappeared precisely because vaccines exist; however, the bacteria and viruses that cause them have not disappeared. In Venezuela, where no cases of measles or diphtheria had been reported for decades, 7,524 suspected cases of measles and 2,170 cases of diphtheria were recorded from June 2017 to October 2018, the latter with a mortality rate of 22 percent. The cause was disruptions to the national immunization program, which led to a decrease in vaccination coverage for both diseases by almost 50 percent when the country's public health system imploded as a result of regime mismanagement. Something similar happened in Russia in the 1990s, when diphtheria reappeared with violent epidemics after three decades of silence, after vaccination programs ceased with the collapse of the Soviet regime.

### 7. COVID-19 vaccines have been developed too quickly to be safe.

No shortcuts have been taken on the safety and efficacy of COVID-19 vaccines, at least in advanced countries such as the European Union and the United

States, where laws for approval are strict. The risks were only financial: thanks to the large amount of money made available by governments, it was possible to proceed very quickly. Each vaccine that reached clinical trials' third and last phase was tested in at least 35,000 volunteers on average. That is about twenty times the number of volunteers normally involved in nonemergency clinical trials. For this reason, the vaccines that have been approved for emergency or conditional use against COVID-19 are safe and effective.

# Maybe You Did Not Know That ...

**Microbes have immense power to devastate civilizations as we know them.**

The loss of human life and the social and economic damages caused by COVID-19 are probably incalculable in full, as indeed those of previous pandemics. In 1348 the plague in a few days wiped out Siena, a thriving city with its exuberant economy and culture. The "Spanish" flu of 1918 killed most likely more than 40 million people in less than a year. The AIDS epidemic has been raging for over 35 years, having caused about 36 million deaths and 73 million infections to date: in Africa, AIDS still leaves millions of orphaned children today, who grow up without that traditions, knowledge, and teachings are transmitted to them from the previous generation, having been wiped out by the disease.

**Vaccines are an extraordinary reason for optimism about our abilities to defend ourselves against infections.**

The increase in the life expectancy in most western countries, now over eighty years of age, depends significantly on the fact that (almost) everyone is vaccinated.

Microbes are, in a way, far ahead of us in evolutionary abilities: they adapt very quickly to every new opportunity and keep challenging our immune system. But we too have made extraordinary progress, by increasing our defense capabilities in ways that were unthinkable until a few years ago.

**Vaccines help our defenses do a better job.**

A vaccine is like a flight simulator for a pilot. Just as the inexperienced pilot avoids putting passengers at risk, the vaccine allows the immune system to experience what can happen during an illness, without the body having to suffer the negative consequences. In fact, the vaccine contains the instructions for identifying the pathogen and the ways in which it can be combated, just as the flight simulator contains a good sample of what a pilot may encounter during his flights.

**Vaccines are far more effective than the best existing medicines.**
They cost little compared to other treatments; they save billions of human lives; and they are more effective than the best medicines (the most effective medicines are those against the rheumatic fever and the rheumatic heart disease, with an efficacy of 75 percent, versus an approximate 98 percent efficacy for routine vaccines).

**In poor countries, vaccines are the most efficient way to "buy" health and prosperity.**
For every $1 spent on vaccinations in the ninety-four lowest-income countries in the world, savings are expected to save $16 in healthcare costs, wages, and lost productivity due to illness and death, according to a study conducted at the Johns Hopkins University. If you include broader benefits, such as the value people place on healthier, longer lives, and the long-term burdens of disabilities, the net return increases to $44 per dollar invested.

**More than one technological revolution is underway in the art of building vaccines.**
In the current postgenomic era, vaccines are no longer raw preparations: they have become safer and more effective compounds, designed through very rapid computer simulations after the gene sequence of a new virus or bacterium is shared on the internet. The increasingly sophisticated immunological knowledge, together with the progress made by structural vaccinology, allows scientists to select with great precision the antigen to be included in the vaccine. In the most advanced preparations, the antigen can be built by our cells, as long as they are provided with instructions in the form of synthetic mRNA wrapped in a nanolipid envelope. The antigen, or the instructions for building it, is sometimes modified to improve the immune response, which can be further induced by the addition of an effective adjuvant. With these new technologies, diseases such as meningitis and flu are now more preventable, even though their germs are masters of disguises and were considered beyond reach

with vaccines. Also, thanks to technological innovations, the time it takes to discover and bring a new vaccine to clinical trials and approval has been dramatically reduced: where it once took about fifteen years, today it can take less than a year, as we have seen with the COVID-19 vaccines.

**Thanks to the prevention of childhood diseases around the world, more and more people are enjoying a long life with their children and grandchildren.**

In 1990, the age group with the highest mortality rate was by far that of children under 5 years of age. In 2017, the number of deaths among the youngest had halved and the 80–84 years old had become the age group with the highest mortality rate. The credit for this improvement goes, among others, to Gavi, the Vaccine Alliance, which supplies vaccines to countries that cannot afford them.

**Any new vaccine arrives in developing countries with an average delay of a few decades, compared to the richest nations.**

It can take several years before the price of a new vaccine becomes affordable for the poorest countries. Because of the time gap, many avoidable diseases and deaths still happen.

**An international alliance is increasing access to vaccinations in the poorest countries.**

Active in 73 countries, Gavi, the Vaccine Alliance, is a public-private partnership whose mission is to improve the health of those living in developing countries by aggregating the demand for vaccines and obtaining lower prices from manufacturers. From 2000–2020, Gavi contributed to the routine vaccination of 883 million children, about half of the world's, preventing over 15 million deaths.

**Most of the vaccines used in the world are produced in just nine countries.**

The ability to design and produce vaccines is an important strategic and economic resource for any country. Nations that do not have such a production and research capacity are forced to import vaccines from abroad.

**Vaccine safety: A challenge for the mind.**

The vaccinations of the twenty-first century are safer, smarter, and more effective than those available in the twentieth century. This progress in vaccine research has occurred very quickly compared to our way of thinking, which sometimes struggles in the face of novelty. It is time for our mental "hard drive" to update, because there are unfounded fears to abandon, and excellent opportunities for our health to be seized.

# Further Reading

A selection of books, articles, and websites for those who want to learn more about vaccines, infectious diseases, and immunity.

## Old and new epidemics

Burdick, Alan. "Monster or Machine? A Profile of the Coronavirus at 6 Months." *The New York Times*, June 2, 2020. https://www.nytimes.com/2020/06/02/health/coronavirus-profile-covid.html.

Crawford, Dorothy. "Viruses: The Invisible Enemy Revealed." The Future of Science. September 19–21, 2010. https://www.thefutureofscience.org/uploads/63d66469bd95dfe09829ae708540025f.pdf.

Dennett, Daniel. *Consciousness Explained*. United Kingdom: Little, Brown and Co, 1991.

Kolbert, Elizabeth. "Pandemics and the Shape of Human History." *The New Yorker*, March 30, 2020. https://www.newyorker.com/magazine/2020/04/06/pandemics-and-the-shape-of-human-history.

Meacham, Jon. "Pandemics of the Past." *The New York Times*, May 7, 2020. https://www.nytimes.com/2020/05/07/books/review/pandemics-of-the-past-coronavirus.html.

Our World in Data. https://ourworldindata.org/.

The World Bank. "Life expectancy at birth (total years)—Sub-Saharan Africa." https://data.worldbank.org/indicator/SP.DYN.LE00.IN?end=2017&locations=ZG&name_desc=true&start=1960&view=chart.

UNAIDS. "Botswana." https://www.unaids.org/en/regionscountries/countries/botswana.

Vozza, Lisa. "Small notes scattered on the margins of viral epidemics." Aula di Scienze, Zanichelli. March 2, 2020, https://aulascienze.scuola.zanichelli.it/biologia-e-dintorni/2020/03/02/piccole-note-sparse-a-margine-di-epidemie-virali/.

World Health Organization. "HIV/AIDS." July 17, 2021. https://www.who.int/newsroom/fact-sheets/detail/hiv-aids.

## The long history of vaccines

Barquet, Nicolau, and Pere Domingo. "Smallpox: The Triumph over the Most Terrible of the Ministers of Death." *Annals of Internal Medicine* 127, no. 8 Part 1 (1997): 635–42. https://doi.org/10.7326/0003-4819-127-8_part_1-199710150-00010.

Conis, Elena, Michael Mccoyd, and Jessie A. Moravek. "What to Expect When a Coronavirus Vaccine Finally Arrives." *The New York Times*, May 20, 2020. https://www.nytimes.com/2020/05/20/opinion/coronavirus-vaccine-polio.html.

Fredericks, David N., and David A. Relman. "Sequence-Based Identification of Microbial Pathogens: A Reconsideration of Koch's Postulates." *Clinical Microbiology Reviews* 9, no. 1 (1996): 18–33. https://doi.org/10.1128/cmr.9.1.18.

Klass, Perri. "Hoping for a Covid Vaccine and Recalling the One for Smallpox." *The New York Times*, May 25, 2020. https://www.nytimes.com/2020/05/25/well/family/covid-vaccine-smallpox-coronavirus.html.

Lipkin, W. Ian. "Microbe Hunting in the 21st Century." *PNAS* 106, no 1 (2009): 6–7. https://doi.org/10.1073/pnas.0811420106.

Loomis, Joshua S. *Epidemics: The Impact of Germs and Their Power Over Humanity.* California: Praeger, 2018.

Racaniello, Vincent. "Koch's postulates in the 21st century." Virology Blog. January 22, 2010. https://www.virology.ws/2010/01/22/kochs-postulates-in-the-21st-century/.

Racaniello, Vincent. "Leaving Koch behind." Virology Blog. December 15, 2018. https://www.virology.ws/2018/11/15/leaving-koch-behind/.

Thucydides, and Rex Warner. *History of the Peloponnesian War.* Baltimore: Penguin books, 1963.

Treccani. "Redi, Francesco." http://www.treccani.it/enciclopedia/francesco-redi_%28Dizionario-di-Medicina%29/.

Treccani. "Spallanzani, Lazzaro." http://www.treccani.it/enciclopedia/lazzaro-spallanzani/.

Willis, N J. "Edward Jenner and the Eradication of Smallpox." *Scottish Medical Journal* 42, no. 4 (1997): 118–21. https://doi.org/10.1177/003693309704200407.

World Health Organization. "Poliomyelitis." https://www.who.int/news-room/fact-sheets/detail/poliomyelitis.

## Listen to this

American Academy of Pediatrics. "Vaccine Safety: Examine the Evidence." Healthy Children.org. July 7, 2018. https://www.healthychildren.org/English/safety-prevention/immunizations/Pages/Vaccine-Studies-Examine-the-Evidence.aspx.

Autismoatreviso. "Autismo Treviso Onlus." http://www.autismotreviso.org/.

Barbash, Fred. "The saddest story Roald Dahl ever wrote—about his daughter's death from measles—is worth reading today." *Washington Post*, February 2, 2015. https://www.washingtonpost.com/news/morning-mix/wp/2015/02/02/the-saddest-story-roald-dahl-ever-wrote-about-his-daughters-death-from-measles-and-is-worth-reading-today/.

Broad, William J. "Putin's Long War Against American Science." *The New York Times*, April 13, 2020. https://www.nytimes.com/2020/04/13/science/putin-russia-disinformation-health-coronavirus.html.

Chernikoff, Leah. "When a Close Friend Has Doubts About Vaccinations." *The New York Times*, May 28, 2020. https://www.nytimes.com/2020/05/28/parenting/coronavirus-vaccine-parents.html.

European Society of Clinical Microbiology and Infectious Diseases. "Vaccine-preventable diseases surge in crisis-hit Venezuela." EurekAlert AAAS. April 15, 2020. https://www.eurekalert.org/pub_releases/2019-04/esoc-vds041219.php.

Gross, Liza. "A Broken Trust: Lessons from the Vaccine–Autism Wars." *PLoS Biology* 7, no. 5 (2009): 1–7. https://doi.org/10.1371/journal.pbio.1000114.

Hoffman, Jan. "One More Time, with Big Data: Measles Vaccine Doesn't Cause Autism." *The New York Times*, March 5, 2019. https://www.nytimes.com/2019/03/05/health/measles-vaccine-autism.html.

Mikulic, Matej. "People Who Believed Vaccines Cause Autism Worldwide 2017." *Statista*. December 8, 2017. https://www.statista.com/statistics/790868/people-who-believed-vaccines-cause-autism-worldwide-select-countries/.

Mohney, Gillian. "7-Year-Old Cancer Patient Makes Adorable Argument for Vaccines." *ABC News*, February 12, 2015. https://abcnews.go.com/Health/boy-battled-leukemia-lobbies-vaccinated-students/story?id=28919133.

Offit, Paul A. *Autism's False Prophets Bad Science, Risky Medicine, and the Search for a Cure*. New York: Columbia University Press, 2010.

Parker, Amy. "Growing Up Unvaccinated." *Voices for Vaccines*. December 9, 2013. https://www.voicesforvaccines.org/growing-up-unvaccinated/.

Racaniello, Vincent. "An advocate for vaccines." Virology Blog. April 11, 2019, https://www.virology.ws/2019/04/11/an-advocate-for-vaccines/.

Royal Society for the Prevention of Accidents. "Home Accident Statistics in Scotland." https://www.rospa.com/home-safety/uk/scotland/research/statistics.

Starr, Douglas. "This Italian scientist has become a celebrity by fighting vaccine skeptics." *Science*, January 1, 2020. https://www.sciencemag.org/news/2020/01/italian-scientist-has-become-celebrity-fighting-vaccine-skeptics.

Statista Research Department. "Topic: Road Safety in the UK." *Statista*. October 20, 2021. https://www.statista.com/topics/5255/road-safety-in-the-uk/#:~:text= Preliminary%20data%20for%202018%20indicates,Great%20Britain%20 amounted%20to%20165%2C050.

The Children's Hospital of Philadelphia. "Autism, Unlocked." July 17, 2018. https:// www.chop.edu/news/autism-unlocked.

UK Health Security Agency. "Confirmed cases of measles, mumps and rubella in England and Wales: 1996 to 2020." Gov.uk. June 21, 2021. https://www.gov.uk/govern-ment/publications/measles-confirmed-cases/confirmed-cases-of-measles-mumps-and-rubella-in-england-and-wales-2012-to-2013.

World Health Organization. "Autism spectrum disorders." https://www.who.int/news-room/fact-sheets/detail/autism-spectrum-disorders.

World Health Organization. "New Measles Surveillance Data for 2019." May 15, 2019. https://www.who.int/news/item/15-05-2019-new-measles-surveillance-data-for-2019.

World Health Organization. "Vaccine-preventable diseases: monitoring system. 2019 global summary." July 15, 2020. https://apps.who.int/immunization_monitoring/globalsummary/countries?countrycriteria%5Bcountry%5D%5B%5D=GBR&commit=OK.

## Chasing moving targets

Abbott, Alison. "A Place in the Sun." *Nature* 446, no. 7132 (2007): 124–25. https://doi.org/10.1038/446124a.

Bebe Vio. http://www.bebevio.com/.

Centers for Disease Control and Prevention. "Meningococcal Disease." May 31, 2019. https://www.cdc.gov/meningococcal/about/causes-transmission.html.

Centers for Disease Control and Prevention. "Pertussis (Whooping Cough)." https://www.cdc.gov/pertussis/countries/index.html.

Centers for Disease Control and Prevention. "Respiratory Syncytial Virus Infection (RSV)." December 18, 2020. https://www.cdc.gov/rsv/index.html.

Centers for Disease Control and Prevention. "Vaccines During and After Pregnancy." November 9, 2021. https://www.cdc.gov/vaccines/pregnancy/vacc-during-after.html.

Choo, Qui-Lim, George Kuo, Amy J. Weiner, Lacy R. Overby, Daniel W. Bradley, and Michael Houghton. "Isolation of a cDNA Clone Derived from a Blood-Borne Non-A, Non-B Viral Hepatitis Genome." *Science* 244, no. 4902 (1989): 359–62. https://doi.org/10.1126/science.2523562.

Crank, Michelle C., Tracy J. Ruckwardt, Man Chen, and Kaitlyn M. Morabito, Emily Phung, Pamela J. Costner, LaSonji A. Holman, *et al.* "A Proof of Concept for Structure-Based Vaccine Design Targeting RSV in Humans." *Science* 365, no. 6452 (2019): 505–9. https://doi.org/10.1126/science.aav9033.

Inserm Press Office. "Hepatitis C virus observed under a microscope for the first time." *Inserm.* October 19, 2016. https://presse.inserm.fr/en/hepatitis-c-virus-observed-under-a-microscope-for-the-first-time/25440/.

Los Alamos National Laboratory. "HIV sequence database." September 7, 2021. https://www.hiv.lanl.gov/content/sequence/HIV/mainpage.html.

Mascola, John. "Structure-Based Vaccine Design and B-cell Ontogeny in the Modern Era of Vaccinology." *YouTube.* December 9, 2016. https://youtu.be/lzQCaZbqNAM.

Medini, Duccio, Davide Serruto, Julian Parkhill, David A. Relman, Claudio Donati, Richard Moxon, Stanley Falkow, *et al.* "Microbiology in the Post-Genomic Era." *Nature Reviews Microbiology* 6, no. 6 (2008): 419–30. https://doi.org/10.1038/nrmicro1901.

National Institute of Allergy and Infectious Diseases. "Study offers clues to making vaccine for infant respiratory illness." *MedicalXpress.* April 25, 2013. https://medicalxpress.com/news/2013-04-clues-vaccine-infant-respiratory-illness.html.

National Institutes of Health. "Genome Information by Organism." *National Center for Biotechnology Information.* https://www.ncbi.nlm.nih.gov/genome/browse#!/overview/.

NobelPrize.org. "The Nobel Prize in Chemistry 2020." https://www.nobelprize.org/prizes/chemistry/2020/summary/.

NobelPrize.org. "The Nobel Prize in Physiology or Medicine 1984." https://www.nobelprize.org/prizes/medicine/1984/summary/.

Oliver, Sarah E. "Manual for the Surveillance of Vaccine-Preventable Diseases, Chapter 2: Haemophilus influenzae invasive disease." *Centers for Disease Control and Prevention.* April 1, 2014. https://www.cdc.gov/vaccines/pubs/surv-manual/chpt02-hib.html.

Rappuoli, Rino. "From Pasteur to Genomics: Progress and Challenges in Infectious Diseases." *Nature Medicine* 10, no. 11 (2004): 1177–85. https://doi.org/10.1038/nm1129.

Telford, John L. "Bacterial Genome Variability and Its Impact on Vaccine Design." *Cell Host & Microbe* 3, no. 6 (2008): 408–16. https://doi.org/10.1016/j.chom.2008.05.004.

US Food and Drug Administration. "FDA News Release: First vaccine approved by FDA to prevent serogroup B Meningococcal disease." October 29, 2014. https://www.fda.gov/news-events/press-announcements/first-vaccine-approved-fda-prevent-serogroup-b-meningococcal-disease.

Wikipedia. "William Rutter." June 18, 2021. https://en.wikipedia.org/wiki/William_J._Rutter.

World Health Organization. "Haemophilus Influenzae type b (Hib)." https://www.who.int/teams/health-product-policy-and-standards/standards-and-specifications/vaccine-standardization/hib.

World Health Organization. "Hepatitis B." July 27, 2021. https://www.who.int/news-room/fact-sheets/detail/hepatitis-b.

World Health Organization. "Hepatitis Vaccine." http://www.emro.who.int/health-topics/hepatitis-vaccine/.

World Health Organization. "Pertussis." https://www.who.int/health-topics/pertussis#tab=tab_1.

Yeung, Karene Hoi, Philippe Duclos, E. Anthony Nelson, and Raymond Christiaan W. Hutubessy. "An Update of the Global Burden of Pertussis in Children Younger than 5 Years: A Modelling Study." *The Lancet Infectious Diseases* 17, no. 9 (2017): 974–80. https://doi.org/10.1016/s1473-3099(17)30390-0.

## Vaccines wanted against four big killers

Cohen, Jon. "Another HIV vaccine strategy fails in large-scale study." *Science*, February 3, 2020. https://www.sciencemag.org/news/2020/02/another-hiv-vaccine-strategy-fails-large-scale-study.

Duffy, Patrick E., and J. Patrick Gorres. "Malaria Vaccines since 2000: Progress, Priorities, Products." *npj Vaccines* 5, no. 1 (2020). https://doi.org/10.1038/s41541-020-0196-3.

Herchline, Thomas E. "What is the global prevalence of tuberculosis (TB)?" *Medscape*. June 4, 2020. https://www.medscape.com/answers/230802-19522/what-is-the-global-prevalence-of-tuberculosis-tb.

Johnston, Margaret I., and Anthony S. Fauci. "An HIV Vaccine—Challenges and Prospects." *New England Journal of Medicine* 359, no. 9 (2008): 888–90. https://doi.org/10.1056/nejmp0806162.

Kaufmann, Stefan H. "Deadly Combination." *Nature* 453, no. 7193 (2008): 295–96. https://doi.org/10.1038/453295a.

Leonard, Kimberly. "Tuberculosis Caused More Deaths than HIV in 2014." *US News*, October 28, 2015. https://www.usnews.com/news/articles/2015/10/28/tuberculosis-passes-hiv-as-no-1-infectious-disease.

Malaria Vaccine Initiative. https://www.malariavaccine.org/.

McNeil, Donald G. Jr. "New TB Vaccine Could Save Millions of Lives, Study Suggests." *The New York Times*, October 29, 2019. https://www.nytimes.com/2019/10/29/health/tuberculosis-vaccine.html.

National Center for Biotechnology Information. "Genomes." https://www.ncbi.nlm.nih.gov/genome.

NobelPrize.org. "The Nobel Prize in Physiology or Medicine 2020." https://www.nobelprize.org/prizes/medicine/2020/summary/.

Rabinovich, N. Regina. "Are We There Yet? The Road to a Malaria Vaccine." *The Western Journal of Medicine* 176, no. 2 (2002): 82–84. https://www.ncbi.nlm.nih.gov/pmc/articles/PMC1071669/.

TB Alliance. https://www.tballiance.org/.

The Scripps Research Institute. "Scientists uncover why hepatitis C virus vaccine has been difficult to make." *Science Daily*. October 24, 2016. https://www.sciencedaily.com/releases/2016/10/161024161651.htm.

US Military HIV Research Program. "Frequently asked questions regarding the RV144 Prime-Boost HIV Vaccine Trial." *HIVresearch.org*. March 25, 2010. https://www.who.int/immunization/sage/RV144_FAQs.pdf?ua=1.

World Health Organization. "BCG Vaccine." https://www.who.int/biologicals/areas/vaccines/bcg/en/.

World Health Organization. "GSK's Investigational Vaccine Candidate M72/AS01E shows promise for prevention of TB disease in a Phase 2b trial conducted in Kenya, South Africa and Zambia." September 25, 2018. https://www.who.int/news/item/25-09-2018-gsk-s-investigational-vaccine-candidate-m72-as01e-shows-promise-for-prevention-of-tb-disease-in-a-phase-2b-trial-conducted-in-kenya-south-africa-and-zambia.

World Health Organization. "Hepatitis C." July 27, 2021. https://www.who.int/news-room/fact-sheets/detail/hepatitis-c.

World Health Organization. "Malaria." October 28, 2021. https://www.who.int/news-room/fact-sheets/detail/malaria.

World Health Organization. "Tuberculosis." October 14, 2021. https://www.who.int/en/news-room/fact-sheets/detail/tuberculosis.

World Health Organization. "World Hepatitis Day 2019." https://www.who.int/campaigns/world-hepatitis-day/2019.

World Population Review. "Population of Cities in United Kingdom (2021)." https://worldpopulationreview.com/countries/cities/united-kingdom.

UNAIDS. "Global HIV & AIDS statistics—Fact sheet." https://www.unaids.org/en/resources/fact-sheet.

Zingaretti, C., R. De Francesco, and S. Abrignani. "Why is it so difficult to develop a hepatitis C virus preventive vaccine?" *Clinical Microbiology and Infection* 20 (2014): 103–9. https://doi.org/10.1111/1469-0691.12493.

## Unpredictable viruses: The flu

Centers for Disease Control and Prevention. "Adjuvants and Vaccines." August 14, 2020. https://www.cdc.gov/vaccinesafety/concerns/adjuvants.html.

Centers for Disease Control and Prevention. "How Influenza (Flu) Vaccines Are Made." August 31, 2021. https://www.cdc.gov/flu/prevent/how-fluvaccine-made.htm.

Centers for Disease Control and Prevention. "Influenza A Subtypes and the Species Affected." September 27, 2018. https://www.cdc.gov/flu/other/animal-flu.html.

Centers for Disease Control and Prevention. "Types of Influenza Viruses." November 2, 2021. https://www.cdc.gov/flu/about/viruses/types.htm.

Haelle, Tara. "Why You Need the Flu Shot Every Year." *The New York Times,* December 12, 2017. https://www.nytimes.com/2017/12/12/smarter-living/why-you-need-the-flu-shot-every-year.html.

Kolata, Gina. "How Pandemics End." *The New York Times*, May 10, 2020. https://www.nytimes.com/2020/05/10/health/coronavirus-plague-pandemic-history.html.

Palese, Peter. "Influenza: Old and New Threats." *Nature Medicine* 10, no. S12 (2004): S82–S87. https://doi.org/10.1038/nm1141.

Public Health England. "Surveillance of Influenza and Other Respiratory Viruses in the UK—Winter 2019 to 2020." Gov.uk. June 2020. https://assets.publishing.service.gov.uk/government/uploads/system/uploads/attachment_data/file/895233/Surveillance_Influenza_and_other_respiratory_viruses_in_the_UK_2019_to_2020_FINAL.pdf.

Rappuoli, Rino, and Giuseppe Del Giudice. *Influenza Vaccines for the Future.* Basel: Birkhäuser, 2008.

Rappuoli, Rino, and Philip R. Dormitzer. "Influenza: Options to Improve Pandemic Preparation." *Science* 336, no. 6088 (2012): 1531–33. https://doi.org/10.1126/science.1221466.

Sautto, Giuseppe A., Greg A. Kirchenbaum, and Ted M. Ross. "Towards a Universal Influenza Vaccine: Different Approaches for One Goal." *Virology Journal* 15, no. 1 (2018). https://doi.org/10.1186/s12985-017-0918-y.

Stöhr, Klaus. "Perspective: Ill Prepared for a Pandemic." *Nature* 507, no. 7490 (2014): S20–S21. https://doi.org/10.1038/507s20a.

US Food and Drug Administration. "FDA's Critical Role in Ensuring Supply of Influenza Vaccine." September 28, 2020. https://www.fda.gov/consumers/consumer-up-dates/fdas-critical-role-ensuring-supply-influenza-vaccine.

Wang, Taia T., and Peter Palese. "Unraveling the Mystery of Swine Influenza Virus." *Cell* 137, no. 6 (2009): 983–85. https://doi.org/10.1016/j.cell.2009.05.032.

World Health Organization—Europe Regional Office. "Surveillance and lab network." https://www.euro.who.int/en/health-topics/communicable-diseases/influenza/seasonal-influenza/surveillance-and-lab-network.

World Health Organization. "Influenza (seasonal)." November 6, 2018. https://www.who.int/news-room/fact-sheets/detail/influenza-(seasonal).

Wu, Katherine J. "Coronavirus Vaccine Makers Are Not Mass-Slaughtering Sharks." *The New York Times*, October 13, 2020. https://www.nytimes.com/2020/10/13/science/sharks-vaccines-covid-squalene.html.

## A pandemic of our time

Adam, David. "The Pandemic's True Death Toll: Millions more than Official Counts." *Nature*, January 20, 2022. https://www.nature.com/articles/d41586-022-00104-8.

Buckley, Chris, David D. Kirkpatrick, Amy Qin, and Javier C. Hernández. "25 Days That Changed The World: How COVID-19 Slipped China's Grasp." *The New York Times*, December 30, 2020. https://www.nytimes.com/2020/12/30/world/asia/china-coronavirus.html.

Diamond, Jared. *Guns, Germs and Steel—A Short History of Everybody for the Last 13,000 Years.* United States: W. W. Norton, 1997.

European Centre for Disease Prevention and Control. "Timeline of ECDC's response to COVID-19." April 14, 2021. https://www.ecdc.europa.eu/en/covid-19/timeline-ecdc-response.

Faust, Jeremy Samuel, Harlan M. Krumholz, Chengan Du, Katherine Dickerson Mayes, Zhenqiu Lin, Cleavon Gilman, and Rochelle P. Walensky. "All-Cause Excess Mortality and COVID-19–Related Mortality among US Adults Aged 25–44 Years, March–July 2020." *JAMA* 325, no. 8 (2021): 785–87. https://doi.org/10.1001/jama.2020.24243.

Gorman, James, and Carl Zimmer. "Another Group of Scientists Calls for Further Inquiry Into Origins of the Coronavirus." *The New York Times*, May 13, 2021. https://www.nytimes.com/2021/05/13/science/virus-origins-lab-leak-scientists.html.

Grover, Natalie. "Covid Has 'Cut Life Expectancy in England and Wales by a Year'." *The Guardian*, December 10, 2020. https://www.theguardian.com/world/2020/dec/10/covid-life-expectancy-england-wales.

Hessler, Peter. "How China Controlled the Coronavirus." *The New Yorker*, August 17, 2020. https://www.newyorker.com/magazine/2020/08/17/how-china-controlled-the-coronavirus.

Hessler, Peter. "Life on Lockdown in China." *The New Yorker*, March 30, 2020. https://www.newyorker.com/magazine/2020/03/30/life-on-lockdown-in-china.

Hessler, Peter. "Nine Days in Wuhan, the Ground Zero of the Coronavirus Pandemic." *The New Yorker*, October 12, 2020. https://www.newyorker.com/magazine/2020/10/12/nine-days-in-wuhan-the-ground-zero-of-the-coronavirus-pandemic.

Intini, Elisabetta. "June Almeida: Story of the Scientist Who Discovered Coronaviruses." *Focus*. April 21, 2020. https://www.focus.it/scienza/salute/june-almeida-storia-della-scienziata-che-scopri-i-coronavirus.

Jabr, Ferris. "How Humanity Unleashed a Flood of New Diseases." *The New York Times*, June 17, 2020. https://www.nytimes.com/2020/06/17/magazine/animal-disease-covid.html.

Johns Hopkins University. "COVID-19 Dashboard." https://coronavirus.jhu.edu/map.html.

LaStampa. "Covid, WHO: The Number of Deaths in the World Two or Three Times Higher than Official Data." May 21, 2021. https://www.lastampa.it/cronaca/2021/05/21/news/covid-l-oms-il-numero-dei-morti-nel-mondo-due-o-tre-volte-superiore-ai-dati-ufficiali-1.40298641.

Marr, Linsey C. "Yes, the Coronavirus Is in the Air." *The New York Times*, July 30, 2020. https://www.nytimes.com/2020/07/30/opinion/coronavirus-aerosols.html.

Maxmen, Amy, and Smriti Mallapaty. "The COVID lab-leak Hypothesis: What Scientists Do and Don't Know." *Nature*, July 8, 2021. https://www.nature.com/articles/d41586-021-01529-3.

McCarthy, Niall. "The Last Coronavirus-Free Countries On Earth." *Statista*, November 13, 2020. https://www.statista.com/chart/21279/countries-that-have-not-reported-coronavirus-cases/.

Quammen, David. "The Virus, the Bats and Us." *The New York Times*, December 11, 2020. https://www.nytimes.com/2020/12/11/opinion/covid-bats.html.

Rabin, Roni Caryn, and Emily Anthes. "The Virus Is an Airborne Threat, the C.D.C. Acknowledges." *The New York Times*, May 7, 2021. https://www.nytimes.com/2021/05/07/health/coronavirus-airborne-threat.html.

Racaniello, Vincent. "TWiV 641: COVID-19 with Dr. Anthony Fauci." July 18, 2020. https://www.virology.ws/2020/07/18/twiv-641-covid-19-with-dr-anthony-fauci/.

Roberts, Siobhan. "The Swiss Cheese Model of Pandemic Defense." *The New York Times*, December 5, 2020. https://www.nytimes.com/2020/12/05/health/corona virus-swiss-cheese-infection-mackay.html.

World Health Organization. "Update on Clinical Long-Term effects of Covid-19." March 26, 2021. https://www.who.int/docs/default-source/coronaviruse/ risk-comms-updates/update54_clinical_long_term_effects.pdf?sfvrsn=3e 63eee5_8.

Wright, Lawrence. "The Plague Year." *The New Yorker*, December 28, 2021. https:// www.newyorker.com/magazine/2021/01/04/the-plague-year.

Wu, Katherine J. "How Confident Can You Be in a Coronavirus Test?" *The New York Times*, December 12, 2020. https://www.nytimes.com/2020/12/23/upshot/ coronavirus-tests-positives-negatives.html.

## How to vaccinate an entire planet against COVID-19

Akiko Iwasaki, Akiko, and Ruslan Medzhitov. "Scared That Covid-19 Immunity Won't Last? Don't Be." *The New York Times*, July 31, 2020. https://www.nytimes. com/2020/07/31/opinion/coronavirus-antibodies-immunity.html.

Barda, Noam, Noa Dagan, Yatir Ben-Shlomo, Eldad Kepten, Jacob Waxman, Reut Ohana, Miguel A. Hernán, *et al.* "Safety of the BNT162B2 Mrna COVID-19 Vaccine in a Nationwide Setting." *New England Journal of Medicine* 385, no. 12 (2021): 1078–90. https://doi.org/10.1056/nejmoa2110475.

Carroll, Aaron E. "The Risks of the Covid Vaccine, in Context." *The New York Times*, December 30, 2020. https://www.nytimes.com/2020/12/30/opinion/covid-vaccine-allergic-reactions.html.

Castle, Stephen, and Elian Peltier. "After Botched Covid Response, U.K. Tackles Giant Vaccine Rollout." *The New York Times*, December 7, 2020. https://www.nytimes. com/2020/12/07/world/europe/covid-uk-vaccine-pfizer.html.

Corum, Jonathan, and Carl Zimmer. "How Moderna's Vaccine Works." *The New York Times*, May 7, 2021. https://www.nytimes.com/interactive/2020/health/ moderna-covid-19-vaccine.html.

Corum, Jonathan, and Carl Zimmer. "How the Novavax Vaccine Works." *The New York Times*, May 7, 2021. https://www.nytimes.com/interactive/2020/health/ novavax-covid-19-vaccine.html.

Corum, Jonathan, and Carl Zimmer. "How Oxford-AstraZeneca Vaccine Works." *The New York Times*, May 7, 2021. https://www.nytimes.com/interactive/2020/health/oxford-astrazeneca-covid-19-vaccine.html.

Corum, Jonathan, and Carl Zimmer. "How the Sinopharm Vaccine Works." *The New York Times*, August 4, 2021. https://www.nytimes.com/interactive/2020/health/sinopharm-covid-19-vaccine.html.

Dolgin, Elie. "The Tangled History of mRNA Vaccines." *Nature*, September 14, 2020. https://www.nature.com/articles/d41586-021-02483-w.

Forni, Guido. "Covid Vaccine, Why so Early? The 4 Factors of the Great Race." *Huffington Post*, December 29, 2020. https://www.huffingtonpost.it/entry/vaccino-covid-perche-cosi-presto-i-4-fattori-della-grande-corsa-di-g-forni_it_5feb003fc5b66809cb33f20f.

Gelles, David. "The Husband-and-Wife Team Behind the Leading Vaccine to Solve Covid-19." *The New York Times*, November 11, 2020. https://www.nytimes.com/2020/11/10/business/biontech-covid-vaccine.html.

Goldberg, Emma. "Vaccine Memories of Another Time and Place." *The New York Times*, December 12, 2020. https://www.nytimes.com/2020/12/25/health/covid-vaccine-polio.html.

Goldstein, Joseph. "Hospital Workers Start to 'Turn Against Each Other' to Get Vaccine." *The New York Times*, December 24, 2020. https://www.nytimes.com/2020/12/24/nyregion/nyc-hospital-workers-covid-19-vaccine.html.

Grady, Denise. "AstraZeneca Vaccine Carries Slightly Higher Risk of Bleeding Disorders, Study Shows." *The New York Times*, June 9, 2021. https://www.nytimes.com/2021/06/09/health/astrazeneca-vaccine-bleeding-risk-scotland.html.

Hart, Rober. "Fauci Warns Covid Patent Waivers May Not Be Best Way To Help Boost Vaccine Access." *Forbes*, May 4, 2021. https://www.forbes.com/sites/roberthart/2021/05/04/fauci-warns-covid-patent-waivers-may-not-be-best-way-to-help-boost-vaccine-access/.

Hoffman, Jan. "Early Vaccine Doubters Now Show a Willingness to Roll Up Their Sleeves." *The New York Times*, December 16, 2020. https://www.nytimes.com/2020/12/26/health/covid-vaccine-hesitancy.html.

Holder, Josh. "Tracking Coronavirus Vaccinations Around the World." *The New York Times*, November 24, 2021. https://www.nytimes.com/interactive/2021/world/covid-vaccinations-tracker.html.

Ives, Mike. "Dolly Parton, Who Helped Fund the Moderna Vaccine, Gets a 'Dose of Her Own Medicine.'" *The New York Times*, March 2, 2021. https://www.nytimes.com/2021/03/02/world/dolly-parton-moderna-vaccine-covid.html.

Kauffmann, Sylvie. "Europe's Vaccine Rollout Has Descended Into Chaos." *The New York Times*, February 4, 2021. https://www.nytimes.com/2021/02/04/opinion/eu-covid-vaccines.html.

Khamsi, Roxanne. "Pfizer, AstraZeneca … or Both? A Mixed Approach May Hold Promise." *The New York Times*, June 13, 2021. https://www.nytimes.com/2021/06/13/opinion/covid-vaccine-research-mixing.html.

Kormann, Carolyn. "Countdown to a coronavirus vaccine." *The New Yorker*, December 12, 2020. https://www.newyorker.com/magazine/2020/12/14/countdown-to-a-coronavirus-vaccine.

Kremer, Michael, Rino Rappuoli, and Tito Boeri. "Economics, Vaccine Policy and Pandemics." *YouTube*, June 3, 2021. https://www.youtube.com/watch?v=HGPS3JOfCbk&ab_channel=FestivalEconomia.

LaFraniere, Sharon, Katie Thomas, Noah Weiland, David Gelles, Sheryl Gay Stolberg, and Denise Grady. "Politics, Science and the Remarkable Race for a Coronavirus Vaccine." *The New York Times*, November 21, 2020. https://www.nytimes.com/2020/11/21/us/politics/coronavirus-vaccine.html.

McNeil, Donald G. Jr. "My Job? Telling People What Happens Next." *The New York Times*, August 27, 2020. https://www.nytimes.com/2020/08/27/insider/coronavirus-future.html.

MedlinePlus. "Anaphylaxis." November 23, 2021. https://medlineplus.gov/ency/article/000844.htm.

Mwai, Peter. "Coronavirus in Africa: Concern growing over third wave of Covid-19 infections." *BBC Reality Check*, July 16, 2021. https://www.bbc.com/news/world-africa-53181555.

Racaniello, Vincent. "SARS-CoV-2 Reinfection: What Does it Mean?" *Virology Blog*, August 28, 2020. https://www.virology.ws/2020/08/27/sars-cov-2-reinfection-what-does-it-mean/.

Ramzy, Austin. "Chinese Official Acknowledges Low Effectiveness of the Country's Covid Vaccines." *The New York Times*, April 11, 2021. https://www.nytimes.com/2021/04/11/world/china-covid-vaccine.html.

Rappuoli, Rino, Ennio De Gregorio, Giuseppe Del Giudice, Sanjay Phogat, Simone Pecetta, Mariagrazia Pizza, and Emmanuel Hanon. "Vaccinology in the Post–Covid-19 ERA." *Proceedings of the National Academy of Sciences* 118, no. 3 (2021): e2020368118. https://doi.org/10.1073/pnas.2020368118.

Rey, Gertrud U. "Immunity Is Not Binary." *Virology Blog*, June 4, 2020. https://www.virology.ws/2020/06/04/immunity-is-not-binary/.

Robbins, Rebecca, and Jessica Silver-Greenberg. "Vaccination Campaign at Nursing Homes Faces Obstacles and Confusion." *The New York Times*, December 16, 2020. https://www.nytimes.com/2020/12/16/business/covid-coronavirus-vaccine-nursing-homes.html.

Schmidt, Charles. "New COVID Vaccines Need Absurd Amounts of Material and Labor." *Scientific American*, January 4, 2021. https://www.scientificamerican.com/article/new-covid-vaccines-need-absurd-amounts-of-material-and-labor1/.

Twohey, Megan, and Nicholas Kulish. "Bill Gates, the Virus and the Quest to Vaccinate the World." *The New York Times*, November 23, 2020. https://www.nytimes.com/2020/11/23/world/bill-gates-vaccine-coronavirus.html.

Wee, Sui-Lee. "China Gives Unproven Covid-19 Vaccines to Thousands, With Risks Unknown." *The New York Times*, September 26, 2020. https://www.nytimes.com/2020/09/26/business/china-coronavirus-vaccine.html.

Zimmer, Carl. "'This Is All Beyond Stupid.' Experts Worry About Russia's Rushed Vaccine." *The New York Times*, September 26, 2020. https://www.nytimes.com/2020/08/11/health/russia-covid-19-vaccine-safety.html.

Zimmer, Carl. "This New Covid Vaccine Could Bring Hope to the Unvaccinated World." *The New York Times*, May 5, 2021. https://www.nytimes.com/2021/05/05/health/covid-vaccine-curevac.html.

Zimmer, Carl, Jonathan Corum, and Sui-lee Wee. "Coronavirus Vaccine Tracker." *The New York Times*, June 10, 2020. https://www.nytimes.com/interactive/2020/science/coronavirus-vaccine-tracker.html.

## Vaccines for whom?

Bill & Melinda Gates Foundation. https://www.gatesfoundation.org/.

Bloom, David E., Steven Black, David Salisbury, and Rino Rappuoli. "Antimicrobial Resistance and the Role of Vaccines." *Proceedings of the National Academy of Sciences* 115, no. 51 (2018): 12868–71. https://doi.org/10.1073/pnas.1717157115.

Centers for Disease Control and Prevention. "CDC Vaccine Price List." October 18, 2021. https://www.cdc.gov/vaccines/programs/vfc/awardees/vaccine-management/price-list/index.html.

European Medicines Agency. "Shingrix." September 28, 2020. https://www.ema.europa.eu/en/medicines/human/EPAR/shingrix.

Falkow, Stanley. "I Never Met a Microbe I Didn't Like." *Nature Medicine* 14, no. 10 (2008): 1053–57. https://doi.org/10.1038/nm1008-1053.

Gavi, the Vaccine Alliance. https://www.gavi.org/.

Gavi, the Vaccine Alliance. "Cost-effective." September 15, 2020. https://www.gavi. org/vaccineswork/value-vaccination/cost-effective.

Gates, Bill. "Can this Cooler Save Kids from Dying?" *GatesNotes*, October 13, 2018. https://www.gatesnotes.com/Health/The-big-chill.

Gates, Bill. "This Partnership Helped Prevent 13 Million Deaths." *GatesNotes*, July 4, 2020. https://www.gatesnotes.com/Health/Gavi-has-helped-prevent-13-million-deaths.

Gates, Bill. "Why I Think We Can Predict the Future." *GatesNotes*, October 7, 2020. https://www.gatesnotes.com/Health/Professor-Hawking-Fellowship-lecture.

Gates, Bill, and Melinda Gates. "Why We Swing for the Fences—2020 Annual Letter." *GatesNotes*, February 10, 2020. https://www.gatesnotes.com/2020-Annual-Letter.

LesCultures. "Niger." http://www.lescultures.it/wp/francais-cosa-facciamo/cooperazione-presentazione/niger/.

National Cancer Institute. "Cancer Treatment Vaccines." September 24, 2019. https://www.cancer.gov/about-cancer/treatment/types/immunotherapy/cancer-treatment-vaccines#how-do-cancer-treatment-vaccines-work-against-cancer.

Oh, Jason Z., Rajesh Ravindran, Benoit Chassaing, Frederic A. Carvalho, Mohan S. Maddur, Maureen Bower, Paul Hakimpour, *et al.* "TLR5-Mediated Sensing of Gut Microbiota Is Necessary for Antibody Responses to Seasonal Influenza Vaccination." *Immunity* 41, no. 3 (2014): 478–92. https://doi.org/10.1016/j.immuni.2014.08.009.

Rabin, Roni C. "Ask Well: Do I Need the Shingles Vaccine if I've Had Shingles?" *The New York Times—Well Blogs*, April 15, 2016. https://well.blogs.nytimes.com/2016/04/15/ask-well-do-i-need-the-shingles-vaccine-if-ive-had-shingles/.

Rappuoli, Rino, Henry I. Miller, and Stanley Falkow. "The Intangible Value of Vaccination." *Science* 297, no. 5583 (2002): 937–39. https://doi.org/10.1126/science.1075173.

The World Bank. "Immunization at a Glance." https://openknowledge.worldbank.org/bitstream/handle/10986/9794/multi0page.pdf?sequence=1&isAllowed=y.

The World Bank. "Life Expectancy at Birth, Total (Years)—Niger." https://data.worldbank.org/indicator/SP.DYN.LE00.IN?locations=NE.

The World Bank. "Mortality Rate, Under-5 (Per 1,000 Live Births)—Niger." https://data.worldbank.org/indicator/SH.DYN.MORT?locations=NE.

World Health Organization. "Meningococcal Meningitis." September 28, 2021. https://www.who.int/news-room/fact-sheets/detail/meningococcal-meningitis.

## How do our defenses work?

Murphy K, and Casey Weaver. *Jenaway's Immunobiology*. New York: Taylor & Francis, 2017.

NobelPrize.org. "The Nobel Prize in Physiology or Medicine 2018." https://www.nobelprize.org/prizes/medicine/2018/summary/.

Sompayrac, Lauren. *How the Immune System Works*. Hoboken: Wiley-Blackwell, 2019.

Yong, Ed. "Immunology Is Where Intuition Goes to Die." *The Atlantic*, August 5, 2020. https://www.theatlantic.com/health/archive/2020/08/covid-19-immunity-is-the-pandemics-central-mystery/614956/.

# Index